Cereal biotechnology

Related titles from Woodhead's food science, technology and nutrition list:

Kent's technology of cereals (ISBN: 1 85573 361 7)
N L Kent and A D Evers

This well established textbook provides an authoritative and comprehensive study of cereal technology.

Food machinery (ISBN: 1 85573 269 6)
L M Cheng

This book provides a general technical and mechanical background for the basic processing machinery now used for making snacks, baked goods and confectionery. It covers the basic principles, machine design, function, operation and output.

Wheat – Chemistry and utilization (ISBN: 1 56676 348 7)
H J Cornell and A W Hoveling

This book provides the reader with extensive new information on wheat components that will be useful in improving utilization of wheat and the formulation of new and up-graded wheat-based food products.

Details of these books and a complete list of Woodhead's food science, technology and nutrition titles can be obtained by:

- visiting our web site at *www.woodhead-publishing.com*
- contacting Customer Services (e-mail: *sales@woodhead-publishing.com*; fax: +44 (0)1223 893694; tel: +44 (0)1223 891358 ext. 30; address: Woodhead Publishing Ltd, Abington Hall, Abington, Cambridge CB1 6AH, England)

If you would like to receive information on forthcoming titles in this area, please send your address details to: Francis Dodds (address, tel. and fax as above; e-mail: *francisd@woodhead-publishing.com*). Please confirm which subject areas you are interested in.

Cereal biotechnology

Edited by
Peter C Morris and James H Bryce

CRC Press
Boca Raton Boston New York Washington, DC

WOODHEAD PUBLISHING LIMITED
Cambridge England

Published by Woodhead Publishing Limited
Abington Hall, Abington
Cambridge CB1 6AH
England

Published in North and South America by CRC Press LLC
2000 Corporate Blvd, NW
Boca Raton FL 33431
USA

First published 2000, Woodhead Publishing Limited and CRC Press LLC

British Library Cataloguing in Publication Data
A catalogue record for this book is available from the British Library.

Library of Congress Cataloging in Publication Data
A catalog record for this book is available from the Library of Congress.

Woodhead Publishing Limited ISBN 1 85573 498 2
CRC Press ISBN 0 8493 0899 2
CRC Press order number: WP0899

Cover design by The ColourStudio
Project managed by Macfarlane Production Services, Markyate, Hertfordshire
Typeset by MHL Typesetting Limited, Coventry
Printed by T J International, Padstow, Cornwall, England

Contents

Contributors

Chapters 1 and 11

Dr Peter Morris and Dr James Bryce
Department of Biological Sciences
Heriot-Watt University
Riccarton
Edinburgh EH14 4AS
Scotland

Tel: +44 (0) 131 451 3181
Fax: +44 (0) 131 451 3009
E-mail: P.C.Morris@hw.ac.uk

Chapter 2

Dr R Schuurink and Ms J Louwerse
Department of Biological Sciences
Heriot-Watt University
Riccarton
Edinburgh EH14 4AS
Scotland

Tel: +44 (0) 131 449 5111
Fax: +44 (0) 131 451 3009

Chapter 3

Dr M R Davey
Faculty of Science
University Park
Nottingham NG7 2RD

Tel: +44 (0) 115 951 3057
Fax: +44 (0) 115 951 3298
E-mail: mike.davey@nottingham.ac.uk

Chapter 4

Dr David McElroy
Director of Agricultural Business
Operations
Maxygen Inc
515 Galveston Drive
Redwood City
CA 94063
USA

Tel: +1 650 298 5454
Fax: +1 650 364 2715
E-mail: david_mcelroy@maxygen.com

Chapter 5

Professor Robert Henry
Centre for Plant Conservation Genetics
Southern Cross University
PO Box 157
Lismore
NSW
AUSTRALIA

Tel: (02) 6620 3010
Fax: (02) 6622 2080
E-mail: rhenry@scu.edu.au

Chapter 6

Dr Bill Thomas
Scottish Crop Research Institute
Invergowrie
Dundee DD2 5DA
Scotland

Tel: +44 (0) 1382 562731
Fax: +44 (0) 1382 562426
E-mail: wthoma@scri.sari.ac.uk

Chapter 7

Dr Wendy Cooper
PO Box 686
Norwich NR5 0PZ

Tel: +44 (0) 1603 741293
E-mail: wendy.cooper@ukgateway.net

Dr Jermy Sweet
National Institute of Agricultural
Botany
Huntingdon Road
Cambridge CB3 0LE

Tel: +44 (0) 1223 276381
Fax: +44 (0) 1223 277602

Chapter 8

Dr Andrew Lynn
Food Standards & Product
Technology Department
SAC
Auchincruive
Ayr KA6 5HW
Scotland

Tel: +44 (0) 1292 525087
Fax: +44 (0) 1292 525071
E-mail: A.Lynn@au.sac.ac.uk

Chapter 9

Dr Ray Anderson
High Ridge
Marchington
Uttoxeter
Staffs ST14 8LH

Tel: +44 (0) 1283 820333
E-mail:
raymond.anderson@care4free.net

Chapter 10

Dr Eric Evans
Department of Agriculture
University of Newcastle
Newcastle upon Tyne NE1 7RU

Tel: +44 (0) 191 222 6925
Fax: +44 (0) 191 222 7811
E-mail: E.J.Evans@ncl.ac.uk

1

Introduction

P. C. Morris and J. H. Bryce, Heriot-Watt University, Edinburgh

1.1 Cereals: an introduction

Cereals owe their English name to the Roman goddess Ceres, the giver of grain, indicative of the antiquity and importance of cereals (Hill 1937). This importance is still very much the case today; cereals of one sort or another sustain the bulk of mankind's basic nutritional needs, both directly and indirectly as animal feed. It is primarily the grains of cereals that are useful to us, although the vegetative parts of the plant may be used as fodder or for silage production, and straw is used for animal bedding.

Cereals are members of the large monocotyledonous grass family, the *Gramineae*. The flowering organs are carried on a stem called the rachis, which may be branched, and in turn bears spikelets which may carry more than one flower at each node of the rachilla (Fig. 1.1). The spikelets may be organised in a loose panicle as in sorghum, oats and some millets, or in a tight spike, as in wheat. The length of the internodes of the rachis and of the rachilla, and the number of flowers at each node of the spikelet determine the overall architecture. Each spikelet is subtended by two bracts or leaf-like organs termed the glumes, and each flower in the spikelet is enclosed in two bract-like organs called the lemma and palea. The lemma may be extended to form a long awn. In some cereals or cereal varieties the lemma and palea may remain attached to the grain; these are termed hulled or husked grains, such as oats and most barleys, as opposed to naked grains such as most wheats and maize (Fig. 1.1).

The cereals, with the exception of maize, are dioecious. Each flower bears both male organs; the three anthers (six in rice), and female organs; the ovary which carries two feathery stigmas. In maize, the male flowers are borne in

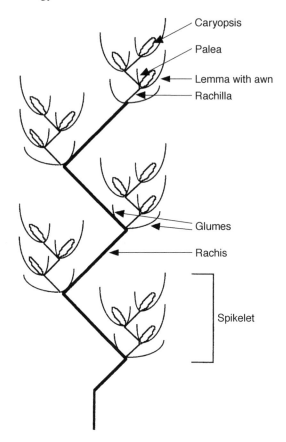

Caryopsis

Palea

Lemma with awn

Rachilla

Glumes

Rachis

Spikelet

Fig. 1.1 Generalised structure of cereal flowering organs. The length and branching pattern of the rachis and the rachilla, and the number of flowers per spikelet determine the overall appearance of the cereal.

spikes on a terminal panicle called a tassel, and the female flowers are in spikelets borne in rows on the swollen tips of lateral branches, the cobs. The main storage organ of cereal seeds is the endosperm which makes up the bulk of the grain, and primarily consists of starch and protein. The grain is botanically a fruit known as a caryopsis; in this structure, the wall of the seed (the testa) becomes fused with the maternally derived ovary wall (the pericarp).

Cereals have developed their importance as food plants because they are high yielding, with world average yields around three tonnes per hectare. The grains are very nutritious; generalised cereal grain contents (which will of course vary with species, growing conditions and variety) are: carbohydrates (70%), protein (10%), lipids (3%) (Pomeranz 1987). Being desiccated at harvest with a water content of about 12%, cereal grains are easy and economical to transport and store. Different cereals have risen to eminence in different quarters of the globe because of geographical provenance and because of differing climatic and

environmental requirements for growth, but their shared favourable characteristics underline their importance as staple foodstuffs. Three cereals – wheat,
maize and rice – make up the bulk of world cereal production, but five other
cereal crops also make important contributions to world nutrition, and to food
and drink production. In order of global production tonnage, these are barley,
sorghum, millet, oats and rye (Fig. 1.2).

1.1.1   Wheat
Wheat is an ancient cultivated crop, whose origins are not clear, but most of the
evidence points towards the Middle East as the geographical region of origin
(DeCandolle 1886, Peterson 1965). There are three sets of wheat species,
differing in ploidy (basic chromosome number). *Triticum monococcum*
(einkorn) is a 'primitive' diploid species (haploid chromosome number 7),
whose use goes back to the Neolithic, and which is still cultivated to some extent
in Europe. *Triticum boeoticum* is a wild form of *T. monococcum*, to be found in
the Balkans and eastern Mediterranean.

Triticum dicoccum (emmer) is a tetraploid wheat (haploid chromosome
number 14), and also an ancient cultivated species, associated with the old
Mediterranean cultures, and still grown in some parts of Europe. It is thought to
be descended from the wild species, *T. dicoccoides*, which is still found in the

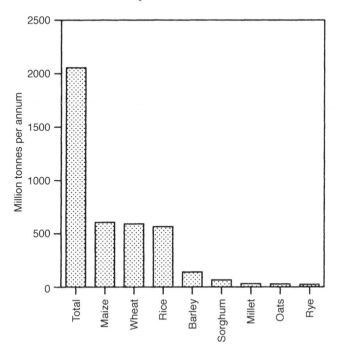

Fig. 1.2   Annual global production of cereals in millions of tonnes (from FAO data for
1998).

eastern Mediterranean region. *Triticum durum* (macaroni wheat), in turn descended from emmer, is grown world wide and has excellent pasta-making qualities. *Triticum timopheevi, T. turgidum* (poulard, rivet or cone wheat), *T. turanicum* (khorasan wheat), *T. polonicum* (Polish wheat or giant rye) and *T. carthlicum* (Persian wheat) are other species of cultivated tetraploid wheat, but now of relatively minor economic importance.

Triticum aestivum is the hexaploid wheat (haploid chromosome number 21) and of all the wheats, this is most commonly grown today. It is thought that diploid einkorn and tetraploid emmer wheats may be ancestral to modern hexaploid wheats. No wild hexaploid species are known, but there are several cultivated subspecies, previously considered by some authorities to be separate

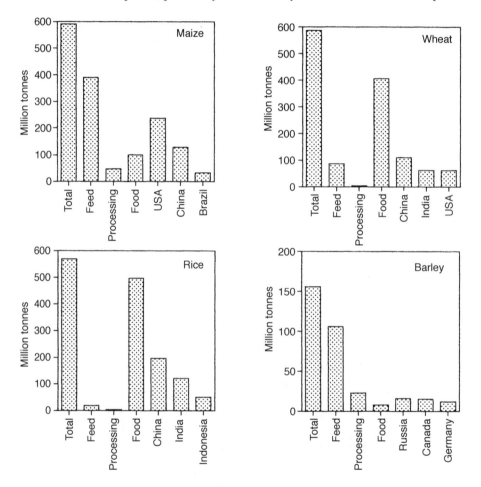

Fig. 1.3   Annual global production and utilisation of the eight most important cereal crops. Total production, utilisation for animal feed, processing (industrial uses and processed foods), and direct human consumption, and the three largest producers are shown (from FAO data for 1996).

species (subsp. *spelta* (spelt or dinkel), *macha, vavilovii, vulgare* (bread wheat), *compactum* (club wheat), and *sphaerococcum* (shot wheat)). The most widespread hexaploid wheat grown today is the bread wheat *Triticum aestivum* subsp. *vulgare*. Winter wheats are sown in autumn, vernalise over winter (vernalisation is a cold treatment required to induce flowering) and are harvested in early summer. Spring wheats are sown in spring and harvested in late summer, they generally have a lower yield than winter wheats (Peterson 1965, Pomeranz 1987).

Wheat is one of the most widely grown cereals, accounting for over one-quarter of the world's global cereal production, and is primarily used for human consumption with some 15% being used for animal feed. The largest global

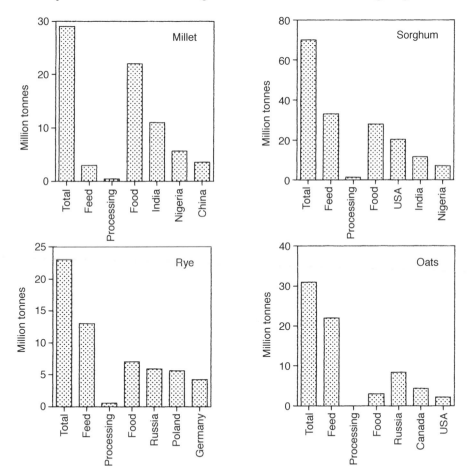

Fig. 1.3   Continued

producers of wheat are China, India and the USA (Fig. 1.3). Wheats can be classified according to kernel hardness: the distinction between hard and soft wheats was made even in Roman times. In American terminology, hard wheats with high protein to starch ratios (16% protein, 61% starch) make 'strong' flour, used in bread-making, whereas 'soft' wheats with 12% protein and 66% starch make 'weak' flours, used in biscuit manufacture. In continental European terminology, 'hard' wheats are durum wheats used for pasta, whilst other wheats are soft (Pomeranz 1987).

1.1.2   Maize

Maize (or corn in North America) (*Zea mays*) derives from and was domesticated in central America some 4000 years ago; a maize goddess, Cinteutl, was worshipped in Mexico (DeCandolle 1886). The true ancestor of maize is not known, but it shares a common ancestor with the weedy species teosinte (*Zea mexicana*). Maize is now grown throughout the world, the main producers being the USA, China and Brazil (Fig. 1.3).

There are many maize subspecies with different agricultural uses, for example varieties *saccharata* (sweetcorn), *everta* (popcorn) *americana* (dent maize, grown in North America), *praecox* (flint maize, grown in Europe), *amylacea* (flour or soft maize, grown by American Indians) and *tunica* (pod corn) (Pomeranz 1987). Maize accounts for over one-quarter of global cereal production, with the majority of the crop going for animal feed; however a substantial tonnage is used directly for human foods and for processing into manufactured foods, drinks and industrial raw materials (Fig. 1.3).

1.1.3   Rice

The most commonly farmed species, *Oryza sativa,* is thought to have been domesticated in southern Asia some 6000 years ago, and written evidence for the cultivation of rice (sometimes termed paddy) in China goes back to at least 2800 BC. Alexander the Great is said to have brought rice to Europe. The progenitor of domesticated rice is the wild species *Oryza rufipogon*. A second rice species (*Oryza glaberrima*) was domesticated in West Africa, and is still an important cultivated species in tropical Africa. Today, rice is a staple foodstuff of Asia and is grown throughout tropical and warm temperate regions. It is grown either immersed in water until harvest (the higher-yielding lowland rice) or on dry land (upland or hill rice). There are two main subspecies of *Oryza sativa*, the generally short-grained *japonica*, typically grown in more northern or southern regions with longer photoperiods, and the longer-grained *indica*, grown in more tropical regions. There are hard- and soft-grained (glutinous) varieties of both subspecies and many thousands of cultivated varieties (Grist 1959, Pomeranz 1987). Rice accounts for over one-quarter of global cereal production, with the vast majority going for human food. China, India and Indonesia account for 65% of the world's production (Fig. 1.3).

1.1.4   Barley

Barley (*Hordeum vulgare*) is of an ancient lineage, being used for bread even before wheat in Neolithic times. It is thought to have arisen in south-western Asia or northern Africa and wild forms of two-row barley are still to be found in western Asia (*Hordeum spontaneum*) (Von Bothmer and Jacobsen 1985, Nevo 1992). Barley is a crop of temperate climates and is a morphologically rather variable species, which has given taxonomists much employment, but in this chapter the view will be taken that there is one cultivated species with several subspecies (for example *distichon, hexastichon, agriocrithon, deficiens*) (Von Bothmer and Jacobsen 1985). Two- and six-row barleys are the most commonly cultivated forms, six-row barley being more resistant to temperature extremes. The spike (or ear) consists of alternating nodes each bearing three spikelets (each a single flower). In two-row barley, only the central spikelet is fertile, but in six-row barley, all three spikelets are fertile. The ear may be erect or drooping at maturity, awned or awn-less. Winter barleys are sown in autumn, vernalised over winter and are harvested in early summer. Spring barleys are sown in spring and harvested in late summer. Barley accounts for some 7.5% of global cereal production (Fig. 1.2). The majority of the barley crop is used as animal feedstuff, but about 15% is used for the production of beer and spirits. Russia, Canada and Germany were the world's biggest barley producers for 1996 (Fig. 1.3).

1.1.5   Oats

Oats have a long and uncertain pedigree, being known since early historical times, for example to the ancient Greeks (DeCandolle 1886). The crop was famously defined by Samuel Johnson in his dictionary as 'a grain, which in England is generally given to horses, but in Scotland supports the people'. However as well as animal food (67% of the world crop), oats are widely used as human nutrition (10%) even outside of Scotland (Fig. 1.3). The most important cultivated species is *Avena sativa*, but other species are also cultivated to a lesser extent, such as *Avena orientalis, A. nuda* and *A. brevis*. There are several wild species of oats, some of which may be ancestral to the cultivated oats, but today are troublesome weeds (for example, *Avena fatua*). Oats form a minor component of the world cereal crop at 1.5% of total global cereal production (Fig. 1.2) and the biggest current producers are Russia, Canada and the USA (Fig. 1.3).

1.1.6   Rye

Rye (*Secale cereale*) appears to be more recently domesticated than other cereals, although it was known to the ancient Greek and Roman civilisations (DeCandolle 1886). The probable ancestor is *Secale montanum*, a wild species to be found in the Black and Caspian Sea areas. Rye is predominantly produced in central and eastern Europe. Both rye and oats may have originated as weed species in wheat and barley crops. Rye is a very winter-hardy crop that will grow on poor soils such as those of the north European plain. Rye accounts for only around 1% of world total cereal production (Fig. 1.2), is used for animal food and human consumption and the prime producers are Russia, Poland and Germany (Fig. 1.3).

1.1.7   Millet

Millet is the collective name for a number of cereal species of importance as food crops in tropical and subtropical countries or as forage crops in more northern climates. These species include: *Eleusine coracana* (finger millet), *Setaria italica* (foxtail millet), *Echinochloa crus-galli* (Japanese barnyard millet), *Pennisetum glaucum* (pearl millet), and *Panicum miliaceum* (proso millet) (Brouk 1975). Although generally low yielding, these crops are often grown in conditions under which other crops would not flourish. Millet forms only 1.5% of the total global cereal crop (Fig. 1.2), and is primarily grown for food. The biggest producers are India, Nigeria and China (Fig. 1.3).

1.1.8   Sorghum

Sorghum (*Sorghum vulgare*) is a native of Africa and Asia and the many varieties which are cultivated there are important as human foods and as animal fodder. There are four general classes, grain sorghum, grass sorghum, broom corn and sweet sorghum or sorgo (Brouk 1975, Pomeranz 1987). Sorghum accounts for 3.5% of global cereal production (Fig. 1.2), the primary producers being the USA, India and Nigeria. In the USA, sorghum is mainly grown for animal fodder, but in India and Nigeria, the majority of the crop is used for human consumption (Fig. 1.3).

1.2   Plant breeding

World cereal production and yield per hectare have increased steadily over the last forty years (Fig. 1.4). This trend is mirrored by the increased use of fertilisers and pesticides (Fig. 1.5). However, much of the increase in yield and production can be attributed to improvements in crop varieties brought about by the efforts of plant breeders.

1.2.1   History of plant breeding

During the course of plant domestication, the elements of plant breeding arose. Early agriculturalists would have taken an empirical approach to selecting their crops. As a result of this, domesticated crops differ from their wild progenitors in a number of important respects. Wild species disperse their seeds in order to spread their offspring far and wide. In wild cereals, the spikes bearing the grains

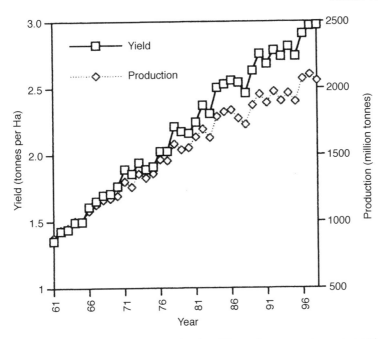

Fig. 1.4   Annual global cereal production in millions of tonnes and cereal yield in tonnes per hectare for the years 1961–98 (from FAO data).

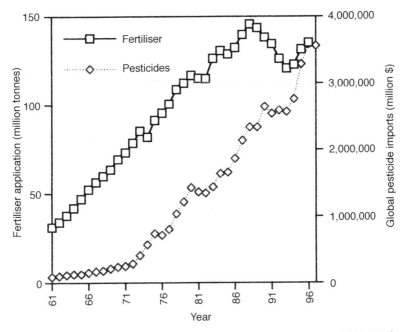

Fig. 1.5   Annual global fertiliser use and pesticide imports over years 1961–97 (from FAO data).

are borne on a brittle rachis (the main axis of the ear on which the grains are carried, Fig. 1.1), and lose the grains if mechanically disturbed. In a cultivated crop, these seeds would be lost prior to harvest, so non-dispersing crops with a rachis that did not easily shatter were 'selected' by man; these crops are threshed after harvest. Domestication would also have put selection pressure on non-dormant crop lines, since only those plants that germinated soon after sowing would be included in the harvest. And of course, crops with enhanced yield and taste would be selected for by the early farmers.

This essentially informal process of selection by early farmers continued right through until recent times, and through time resulted in many 'land races', or plant varieties adapted to local tastes and conditions. In a self-pollinating crop such as barley, these land races (many of which are still in existence today) are composed of many different pure-breeding lines, each of which might have a selective advantage under different environmental pressures. In a cross-breeding crop like maize, land races consist of a genetic continuum with a spectrum of traits across the local population (Chrispeels and Sadava 1994).

Early farmers must have recognised that like begets like, and this allowed for deliberate selection of positive traits, and for some directed crosses to combine these positive traits into one plant. Early records show that date palms were deliberately cross-bred some 5000 years ago. Improvements in crops by breeding depend on two factors; eliminating unwanted characteristics and fostering desired characteristics. Desirable traits can only be selected for if they exist in the local gene pool. Trade and travel would have allowed some limited flux in the gene pools of these early crops. With the coming of global expansion by the European 'superpowers' of the seventeenth century, more and more plant species and varieties became available for farmers to use as crops, and plant breeding was widespread by the early eighteenth century. The recognition that spontaneous mutations or 'sports' could be a source of desirable variation also came about at this time. Notable achievements by the early plant breeders include the crossing of two strawberry species, one from North America and one from Chile, to produce the origin of the modern cultivated strawberry.

The rediscovery of Mendel's laws of inheritance allowed progress to be made in the scientific breeding of crop plants. It was recognised that desirable agricultural features are determined by genetic loci that could be passed on to the offspring of a plant. The genetic mechanisms became understandable and hence more controllable. It became apparent that plants and animals generally have two sets of genes in each cell (the organisms are termed diploid), and that the phenotype, or the characteristics shown by an individual, is a reflection of the expression of these genes. For any given gene, different variants or alleles of that gene exist, which have a dominant or recessive relationship to each other. In a diploid organism, the phenotype of the recessive genes can only be expressed if both copies of the pair of genes in the cell are recessive (the homozygous state), whereas dominant genes give a phenotypic expression even in the presence of one copy of the recessive allele (the heterozygous state).

1.2.2 Modern plant breeding

The object of plant breeding is to improve the quality of the crop. Quality is a subjective term, but might include such traits as yield, flavour, disease or pest resistance, and uniformity. These traits are encoded in the genes that are passed on to the offspring by the parents. The mechanisms of plant breeding are to select for desired attributes within a population, or to introduce traits into that population. Introduced traits might arise within the same species naturally, or may be mutant alleles (spontaneous or induced) of a gene, or may be carried on genes introduced from a species that does not normally breed with the crop species.

Approaches to plant breeding depend on whether the crop is self-pollinating (selfing or in-breeding) or cross-pollinating (or out-breeding). Self-pollinating crops such as wheat, oats, rice and barley have physiological and anatomical mechanisms that ensure that individual flowers are primarily self-fertilising. As a consequence of this, self-pollinating plant populations are composed of individuals, each homozygous for the vast majority, if not all, of the genetic loci, and the progeny of such plants will be identical to the parent, or 'breed true'. In contrast, cross-pollinating plants such as maize exhibit mechanisms to encourage pollen transfer between plants. Cross-pollinating plant populations are composed of individuals with a great degree of genetic heterozygosity. Sexual reproduction by a plant carrying heterozygous genes will result in segregation of alleles in the progeny and consequently a phenotypic segregation; the offspring will be variable in character (Lawrence 1968, Kuckuck *et al.* 1991, Chrispeels and Sadava 1994).

1.2.3 Breeding strategies for in-breeding crops

Selection within in-breeding crops may use single plant or mass selection. An existing mixed population of plants composed of many individuals, each being homozygous, but for differing patterns of alleles at each genetic locus, is subjected to selection for the criteria determined by the breeder. In single plant selection, a large number of individual plants are selected out of the variable population, and compared to each other in subsequent sowings. In mass selection, inferior plants are simply culled.

A technique termed pedigree breeding is the most common method of breeding selfing crops. Pure breeding lines of documented and complementary performance are selected and crossed. The next generation, the F1, will be heterozygous for those loci in which the parents differed. The F1 is self-fertilised and single plant selection takes place in the F2 and subsequent generations. By the F6, after continued self-fertilisation, most lines will be homozygous once more, but each line will have a different pattern of alleles at each variable locus. Other traits, not present in the two original parental lines, can be introduced by crossing them in at the F1 stage of the first cross.

Frequently, an established variety (A) may require improvement by the introduction of only one or a few traits from another variety (B). This can be achieved by back-crossing; making a cross between the two parents (A × B), then back-crossing the F1 to parent A. Selection for the desired character is carried out in the F2, or one generation later (depending on whether the trait is dominant or recessive), and at each subsequent stage before another round of backcrossing. Eventually the progeny of the cross will be homozygous for all the alleles in the recurrent parent A and will contain only the desired trait from B.

1.2.4   Breeding strategies in out-breeding crops

Populations of out-breeding crops share a common gene pool and breeding strategies for these crops are designed to enhance the frequency of favourable genes, and reduce the frequency of disadvantageous genes within that pool. Single plant selection followed by enforced self-fertilisation to ensure homozygosity for favourable traits is often accompanied by a general loss of vigour, termed in-breeding depression. This is attributed to the accumulation in the homozygous state of deleterious recessive genes, normally masked in the heterozygous state.

Mass selection has been a very effective strategy in improving traits such as sugar content of sugar beet, or for oil and protein content in maize. This can be refined by line breeding, which is mass selection followed by single plant selection and subsequent mixing of these lines. The back-cross technique can also be used with out-breeding crops, except that here a small population of plants are used as the recurrent parent.

Another breeding strategy exploits the phenomenon of hybrid vigour. The in-breeding depression caused by homozygosity in out-crossing crops has a corollary, termed heterosis or hybrid vigour. When two in-bred lines are crossed, the F1 generation frequently out-performs the parents. This breeding strategy is often used for commercial crops. In the first instance, homozygous in-bred lines are developed and deleterious traits (infertility, dwarfness, defective seeds, etc.) are removed from the population. In this way, undesirable genes are removed from the line. Continuous selection within lines for normal plants with desirable traits results in homozygous in-bred lines. Crossing two complementary in-bred lines with good general combining ability will give a uniform F1 generation with the attendant advantages of heterosis. Uniformity is an important advantage of the F1 because an out-breeding crop is normally heterogeneous in yield and quality. The skill of the breeder is in ascertaining which two in-bred lines will give an advantageous F1. To aid in the commercial scale production of F1 seed, in-bred lines carrying male sterile genes in one of the parents are used, eliminating the need for hand emasculation. Restorer genes (that restore fertility) in the partner of the cross ensure that the progeny is fertile and will produce the crop. The progeny of the F1 will segregate and revert to a heterogeneous crop, hence the farmer is dependent on the seed company for next year's crop of the same quality. No doubt it would be possible to select similar advantages from an open-pollinated line of maize but for obvious financial reasons, maize breeders have chosen not to do so.

1.2.5   Genetic diversity

Whatever the breeding strategy, an important prerequisite for plant breeding is genetic diversity. Without different genes and alleles of genes, there will be no chance for improvement of our crops. To an extent, genetic diversity exists within the crop varieties that are currently in the fields, and can be induced by application of mutagens. But it is critical that we retain the land races and wild varieties of crop plants that are to be found throughout the world, as a source of genetic variation for the crops of tomorrow (Chrispeels and Sadava 1994).

1.3   Biotechnology: an introduction

Biotechnology is a difficult term to define since the harnessing of any biological process to human aims and desires could justifiably be called biotechnology. However, the revolution in our understanding of the molecular mechanisms underlying the processes of life, in particular our understanding of DNA, the prime genetic material, has resulted in the ability to manipulate those mechanisms to our requirements. This new-found knowledge and ability is loosely termed biotechnology.

There are two main applications of biotechnology to cereals. The first is as an aid to conventional breeding programmes, as outlined above. Physiological or morphological traits are governed by genes carried on chromosomes. The ability to monitor the presence or absence of such genes in plants (even if those genes are in a recessive state or are not otherwise identifiable through the phenotype) is a great aid to plant breeders. This is done through the use of molecular markers, characteristic DNA sequences or fragments that are closely linked to the gene or genes in question. Molecular biological methods allow the monitoring of such markers in many independent individuals, for example those arising from a cross between two cereal varieties. This is a great aid to the selection process (for example Laurie *et al.* 1992).

The second major application of biotechnology is in the ability to transfer genes between different organisms. This means that specific genes can be added to a crop variety in one step, avoiding all the back-crossing that is normally required, providing a major saving of time and effort. Furthermore, those genes that are added need not come from a species that is sexually compatible with the crop in question. Conventional breeding is of course limited to the introduction of genes from plants of the same species or very near relatives. By employing the science of genetic engineering, it is possible to bring into a crop plant, different genes from other plants or even bacteria, fungi or animals. Genes are, simplistically, made up of two parts; the coding region which determines what the gene product is (for example an enzyme like α-amylase, or a seed storage protein like hordein), and the promoter, a set of instructions specifying where, when and to what degree a gene is expressed. Coding regions and promoters from different genes can be spliced together in the laboratory to provide genes with new and useful properties (recombinant DNA). For example, if it were desirable for a heat-stable starch degrading enzyme from a fungus to be expressed during barley germination, the fungal gene could be attached to the promoter of a barley gene that is normally expressed during germination. These foreign or recombinant genes can then be introduced back into crop plants through the techniques of plant genetic transformation. The introduced genes integrate into the plant genome and will be passed on to the offspring in the normal way (Chrispeels and Sadava 1994).

These new approaches to plant breeding are set to revolutionise cereal technology. Already we are seeing the production of crops with properties unimaginable by conventional breeding techniques. We can anticipate cereal crops with improved yields and qualities, and novel, enhanced or optimised properties.

1.4   The structure of this book

We hope that this book will speak to both practising plant molecular biologists, and to those in the cereal-processing industries. This book is not a laboratory cook-book, nor will it be an encyclopaedic work on industrial practice. Rather, we hope to provide an overview of both sides of the coin, to introduce and explain the methods and possibilities of cereal transformation to non-specialists, and likewise to introduce to plant molecular biologists what it is that industrialists actually do with cereals in order to process them, bring them to the market, provide industry with raw materials, and make a profit. Most importantly, we hope to highlight the current limitations to production and processing that could be addressed by molecular biologists. We have brought together leading workers in the field to describe the science behind cereal transformation, concentrating on wheat, barley, rice and maize in Chapters 2 and 3. The commercial development, and production of transgenic cereals and the major traits that can be successfully addressed by this technology, are discussed in Chapters 4 and 5. The use of molecular biology in conventional breeding programmes is discussed in Chapter 6. Chapter 7 deals with the topical and sometimes thorny problems of risk assessment, legislative issues and public perception. Three important chapters (8, 9 and 10) describe current practice and limitations in malting and brewing, milling and baking, and in cereal production, three technology-intensive industries that work with cereals as their prime raw materials.

1.5   Sources of further information and advice

http://www.hgca.com/
The Home-Grown Cereals Authority exists to improve the production and marketing of UK cereals.
http://www.smallgrains.org/Index.htm
A site focusing on the production and marketing of North American cereals.

http://plantbio.berkeley.edu/~outreach/OUTREACH.HTM
An excellent overview of the science and potentials of plant biotechnology.
http://www.fao.org/waicent/search/default.asp
A Food and Agriculture Organisation webpage with links to global agricultural statistics.
http://www.cgiar.org/centers.htm
A Consultative Group on International Agricultural Research website with links to international agricultural research centres including the following:
CIMMYT, International Maize and Wheat Improvement Center.
ICARDA, International Center for Agricultural Research in the Dry Areas (including barley, wheat).
ICRISAT, International Crops Research Institute for Semi-Arid Tropics (including sorghum, millet).
IITA, International Institute of Tropical Agriculture (including maize).
IRRI, International Rice Research Institute.
WARDA, West Africa Rice Development Association.

1.6   References

BROUK B, *Plants Consumed by Man*, London, Academic Press, 1975.

CHRISPEELS MJ and SADAVA DE, *Plants, Genes and Agriculture*, Boston, Jones and Bartlett, 1994.

DeCANDOLLE A, *Origin of Cultivated Plants*, New York, Hafner Publishing Co., 1886 (reprinted 1959).

GRIST DH, *Rice*, London, Longman, 1959.

HILL AF, *Economic Botany*, New York, McGraw-Hill, 1937.

KUCKUCK H, KOBABE G and WENZEL G, *Fundamentals of Plant Breeding*, Berlin, Springer Verlag, 1991.

LAURIE DA, SNAPE JW and GALE MD, 'DNA Marker Technology for Genetic Analysis in Barley'. In: *Barley: Genetics, Biochemistry, Molecular Biology and Biotechnology*, ed. Shewry PR, Oxford, CAB International, 1992, 115–32.

LAWRENCE WJC, *Plant Breeding*, London, Edward Arnold, 1968.

NEVO E, 'Origin, Evolution, Population Genetics and Resources for Breeding of Wild Barley, *Hordeum spontaneum*, in the Fertile Crescent'. In: *Barley: Genetics, Biochemistry, Molecular Biology and Biotechnology,* ed. Shewry PR, Oxford, CAB International, 1992, 19–44.

PETERSON RF, *Wheat*, New York, Interscience Publishers, 1965.

POMERANZ Y, *Modern Cereal Science and Technology*, Weinheim, VCH Publishers, 1987.

VON BOTHMER R and JACOBSEN N, 'Origin, Taxonomy, and Related Species'. In: *Barley*, ed. Rasmusson DC, Madison, American Society of Agronomy, 1985.

2

The genetic transformation of wheat and barley

R. C. Schuurink and J. D. Louwerse, Heriot-Watt University, Edinburgh

The transformation of barley and wheat has become commonplace in the late 1990s. Though transformation procedures are not as routine as for oilseed rape, potato, tomato, maize and rice, several academic institutions and companies have been able to produce transgenic barley and wheat plants. Various patents for transformation procedures as well as many applications of transgenic wheat and barley have been filed. Field trials are being performed suggesting that commercialisation is upon us. Though a three-year moratorium on the commercial growing of transgenic crops has existed since 1999 in the UK, and this moratorium might be extended, the import of transgenic raw materials is not restricted and certainly will affect the cereal biotechnological industries at some point. This chapter aims to explain what is actually meant by transformation, what the transformation of wheat and barley comprises and what are the properties of the current transgenics.

2.1 Introduction

The first reports on the transformation of plants date back more than 15 years now. The first cereal reported to be transformed was rice in the late 1980s, quickly followed by maize and oats in the early 1990s. The first successful transformation of wheat was reported in 1992[1] and a rapid, more commonly used, protocol was published a year later.[2,3] In 1994 three groups reported on the production of transgenic barley plants[4-6] using various methods to be discussed later.

The definition of transformation has varied somewhat over time. This chapter deals with transformation as the stable integration and expression of genetic information which is introduced into wheat and barley by means other than breeding via crosses. In other words, heterologous (derived from a different species) or modified homologous (derived from the same species) genes are introduced into the genetic blueprint (the genome) of the cereal. The cereal will express this new genetic information and the plant will therefore obtain a new phenotype (a new observable characteristic). This new phenotype can be very subtle and might not always be visible to the naked eye. For instance, a wheat plant expressing a new protein in the seed will look identical to a non-transformed wheat plant.

Since the new genetic information is stably integrated, it is implicit that the offspring of the transformants express the introduced genes as well. We will discuss later that occasionally the transgene or the expression of the transgene is lost in later generations. The presence of the transgene can easily be determined at the molecular level. The demonstration of the presence of the transgene at the molecular level is mandatory to be able to call the transformation successful. The transformant needs to be at least partially fertile, i.e. it needs to produce at least healthy pollen or ovules so that offspring can be obtained.

The requirements for obtaining transformants are fourfold:

1. Tissue or cells into which the new genetic information is introduced must be able to regenerate to (partially) fertile plants.
2. Methods to introduce the new genetic information into the cereal cells must be robust.
3. Procedures to identify cells that contain the new genetic information should be selective.
4. The selected cells, which presumably contain the new genetic information, must still be able to regenerate.

It is clear that many different target tissues and different delivery systems of the genetic information have empirically been tried over the years with various results. New target tissues are still being experimented with, although robust protocols for wheat and barley are readily available. These new methods aim to make the transformation protocols less variety dependent so that current commercial varieties can be transformed directly.

Molecular characterisation is always the final proof that indeed transformation has taken place. The standard and accepted procedure is called the genomic Southern hybridisation analysis. This procedure involves the isolation of DNA from the transgenic plants which is separated according to size by means of gel electrophoresis. After gel electrophoresis the DNA is blotted onto a membrane that is subsequently hybridised with the labelled transgene. Only when the transgene is present in the DNA, i.e. on the membrane, will hybridisation occur and will labelled transgene DNA stick to the membrane. This whole procedure takes a few days and provides hard data about the transgenicity of the plants. Other quicker procedures such as the polymerase chain reaction (PCR) are not acceptable as proof for transgenicity. The PCR method can in principle rapidly amplify a transgene from a pool of DNA but it is very difficult, if not impossible, to exclude the presence of false positives. One can also not prove by the PCR method that the transgene is integrated into the genome of the plant.

2.2   Issues in successful transformation

Cereals are commonly considered as difficult to transform, especially wheat and barley. However, reliable transformation protocols do exist and it is anticipated that transformation protocols for cereals will become easier over time. This anticipation is simply based on extrapolation of the situation of other crops that were categorised as recalcitrant to transformation as well. These crops are now, after a lag period, quite easy to transform.

One of the main reasons why it has been so notoriously difficult to transform wheat and barley lies in the fact that there are not as many toti-potent cells present as in for instance tomato and potato plants. A toti-potent cell is defined as a cell that is capable of regenerating to a green fertile plant. As discussed in the next section, the identification of these cells is the most crucial step for a successful transformation. Moreover, the transformation of these toti-potent cells with *Agrobacterium tumefaciens* (see Section 3.1.3), which has been dominantly used in many transformation protocols for other crops, has been successful only for wheat and barley with one cell type present in immature embryos (see Section 3.2.3).

Transformation protocols for wheat and barley were first developed for varieties known to respond favourably in tissue culture. That basically meant that the identified toti-potent cells were able to multiply in culture and could be relatively easily regenerated to a green fertile plant. Application of the same transformation protocols to commercial varieties appeared not to be straight-forward. Two problems occurred: (a) the cells identified as being toti-potent in the model varieties appeared to have lost most of their toti-potency in the commercial varieties; (b) when the cells had retained their toti-potency in the commercial varieties they often multiplied at much longer time intervals than those in the model varieties, therefore requiring very long tissue culture periods. The first problem could sometimes be overcome by using for instance younger immature embryos than for the model variety.[7] The second problem has been approached by changing the tissue culture conditions by varying the medium contents such as plant hormones[8] (see Section 3.2.4) or by adjusting the particle bombardment conditions[9] (see Section 2.4.2). All changes to protocols for model varieties would have to be more or less empirically determined for each commercial variety.

Since data from field trials from the first transformed barley indicate that its agronomic performance (e.g. yield) is less than that of untransformed barley[10] much attention has recently focused on improvement of regenerability and decreased albinism. It quite often occurs during tissue culture that the tissue either loses its regenerability or that the regenerants are albino (literally white, indicating that they have lost their photosynthetic capabilities). It is thought that minimising the tissue culture period, which is necessary to multiply and select the transformed cells before regeneration, will limit this damage to the regenerants and ultimately to the transformants. One can imagine that undesirable subtle changes in the regenerants which are not visible to the naked eye can also occur. These changes might result in reduced agronomic performance. It is thought that these phenotypical changes are due to genetic damage. This means that perhaps the original genetic blueprint is rearranged or modified. It is thought that the length of the tissue culture period and plant hormone regime during the tissue culture phase could damage the genetic information of the cell. New procedures therefore try to steer away from or to minimise the use of synthetic auxin, a plant hormone inducing cell division, which is thought to cause the genetic damage. Including cytokinin, a plant hormone also involved in cell division, seems to improve the regenerability and to decrease the occurrence of albinos.[8,11]

2.3   Target tissues for transformation

The most crucial step for a successful transformation protocol is the identification of cells which can be manipulated *in vitro* (in tissue culture) and which subsequently can be regenerated to a (partially) fertile plant. In other words, the cells of choice have to be toti-potent, or they have to gain this phenotype after the various tissue culture procedures. The chosen cells have to take up the new genetic information, multiply and finally regenerate to a normal plant with reproductive organs. Uptake of genetic information, proliferation and regeneration all show their own efficiencies. One could for instance find a tissue that is almost 10% receptive to foreign genetic material under certain conditions but that only regenerates with an efficiency of 0.001%. In this particular case one might want to search for another target tissue that is more amenable to regeneration. Otherwise one would have to culture an extremely large amount of tissue to obtain only a few regenerants.

For dicotyledons, such as tomato, tobacco and potato, many different tissues and cells have been successfully used for transformation. These protocols quite commonly use *Agrobacterium*, a soil bacterium, to transfer the new genetic information to the plant cells. One procedure for Arabidopsis, a weed used as a model plant in molecular biology, has even abolished the use of tissue culture altogether. It involves the infiltration of *Agrobacterium* into the flowering parts of Arabidopsis by means of vacuum infiltration or wetting agents. The selection for transgenic seeds is done by germinating on a selective medium. Because of the high seed yield of Arabidopsis, even a transformation efficiency of 0.01% can easily result in 50 transgenic seeds. This procedure has not (yet) been extended to other plants though researchers will undoubtedly have experimented in this area. The following section describes the different cells and tissue that have been used for barley and wheat transformation and discusses the advantages and disadvantages of the different protocols.

2.3.1 Protoplasts

Plant cells from which the cell wall is enzymatically removed (Fig. 2.1) are very receptive to the uptake of exogenously provided DNA. Either by a chemical treatment with polyethyleneglycol or an electric treatment method called electroporation (see Section 2.4.1), large amounts of protoplasts can be forced to take up foreign genetic information. Generally, these protoplasts will regenerate their cell walls in a few days when provided with the correct culture medium and will subsequently start to divide. This procedure is routinely used to transform rice and therefore a lot of effort has gone into developing a similar procedure for wheat and barley.[12–18]

To consider protoplasts as the target cells for transformation one first has to decide on the tissue of which the protoplasts are derived. For wheat and barley one of the procedures uses so-called suspension cultures that have a high regeneration capacity.[19,20] These suspension cultures are derived from callus induced on embryos of immature seeds. Callus formation basically involves a tissue culture procedure that deprograms the cells in the immature embryos to become dedifferentiated and therefore in principle toti-potent. Embryo-derived suspension cultures of wheat and barley are normally easy to obtain, but it is relatively seldom that these cultures retain any regeneration capacity. Since transformation frequencies with this procedure are low and the efforts of obtaining regenerable suspension cultures enormous, this procedure has not found wide application in the cereal community. Some research groups use protoplast transformation protocols to evade the patent on the particle gun (see Section 2.4.2).

2.3.2 Microspores

Immature pollen or microspores of wheat and barley can easily be cultured *in vitro* to form embryo-like structures which develop into plants.[21,22] However, as far as we know, only barley microspores have been used successfully to obtain transformants.[5,23,24] Pollen are single cells with a firm cell wall (Fig. 2.2). Barley pollen are haploid (they contain only one set of chromosomes) and the function of mature pollen in the plant is to deliver this set of chromosomes to the ovule during fertilisation. The immature pollen can be triggered by specific tissue culture conditions into embryogenic microspores. Embryo-like structures will appear after a while in these cultures which will develop into green plants when provided with the right conditions. These plants are not diploid, since no fertilisation has taken place, but double haploid. Two sets of identical chromosomes are provided by the microspores, a process that occurs spontaneously in 80% of the microspores. The plants are therefore completely homozygous and this technique is now quite often used in plant breeding to speed up amplification of new varieties.

At first glance microspores seem to be a very good tissue for transformation. However, isolation of microspores from barley is extremely difficult and very genotype dependent. There are two reports describing the successful

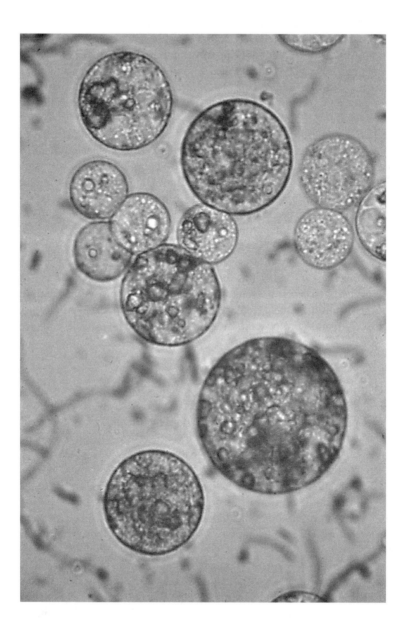

Fig. 2.1 Protoplasts derived from barley suspension-cultured cells. The protoplasts appear round as their cell walls have been removed.

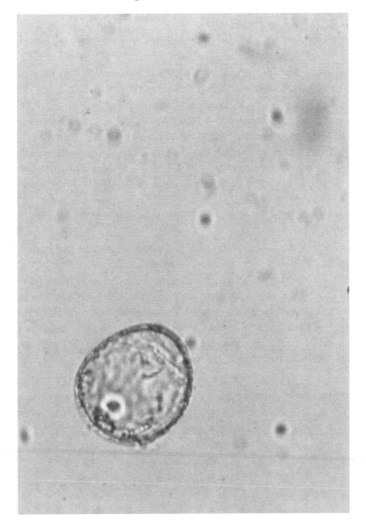

Fig. 2.2   Barley microspore with firm cell wall which is reduced over the germ pore (white spherical structure inside cell).

transformation of barley using microspores.[5,23,24] Both methods used the winter variety Igri and delivery of the DNA was via a particle gun (see Section 2.4.2). The transformation efficiencies were extremely low which might have had something to do with the survival rate of the microspores which were bombarded. It is, however, not unlikely that efficient microspore culture protocols will become available which will renew interest for microspores as a target for transformation.

2.3.3   Immature zygotic embryos

The most commonly used tissue for transformation of wheat and barley has been the immature embryo from the developing grain.[3,6,25–27] The immature embryo is derived from the fertilised ovule which differentiates into an embryo with embryonic root, leaf and a cotyledon (scutellum) after the grain coat has reached a certain size. About 15 to 25 days after the pollen have fertilised the ovules, the immature embryos are isolated (Fig. 2.3). For barley the embryos are cut in half and the scutellum side is used to transfer the new genetic information (the DNA) by means of particle bombardment. Proliferation of the cells resulting in so-called callus formation, is induced by the application of plant hormones after bombardment. This callus is embryogenic and will form green plants after transfer to the right medium. The procedure for wheat is different in the sense that callus formation is induced on the immature embryos prior to particle bombardment.

Though this method was at first limited to certain wheat and barley varieties, considerable effort has now resulted in modified protocols for other cultivars. More than 20 wheat varieties have now been transformed. However, the protocols have to be readjusted for each cultivar and high-quality donor plants for the immature embryos are required. In the meantime it has become clear, in spite of earlier contemplations, that *Agrobacterium* is also capable of transferring DNA to wheat and barley (see Section 2.4.3). The combination of *Agrobacterium* and immature embryos has given rise to higher transformation

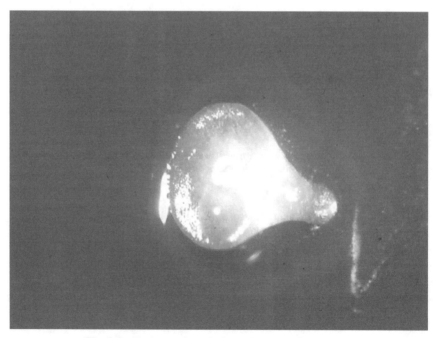

Fig. 2.3   Barley embryo isolated from immature grain.

frequencies for the barley variety Golden Promise, suggesting even higher transformation frequencies in the near future after optimisation.

2.3.4   Apical meristem cultures

The target tissues described so far are all from developing plants thus requiring growth of donor plants under controlled and reproducible conditions. Moreover, the procedures described above involve the extensive use of plant hormone driven tissue culture which can result in abnormal looking regenerants (see also Section 2.2). This so-called somaclonal variation is an unwanted by-product of the transformation procedures and has to be minimised. The transformant should only show a phenotype due to the presence of the new genetic information and not because of the tissue culture. For these reasons, and since all other protocols described are not directly applicable to commercial wheat and barley varieties, some attention has focused on meristematic cultures derived from germinating seeds. This also has the advantage that normal seeds from a field can be used.

Meristemic cultures are initiated by germinating seeds for seven days under sterile conditions. The vegetative shoot which contains the meristem is then isolated and cultured on medium containing very low amounts of auxins. Under these conditions the auxiliary shoots, containing new meristems, will proliferate and these are cut back till the adventitious shoots, containing more meristems, develop. The shoots are cut regularly and a tissue containing just meristems remains (Fig. 2.4). These meristematic tissues form an excellent target for the delivery of new genetic information by particle bombardment. Each meristematic cell can, in principle, give rise to a part of a new meristem and, upon several rounds of meristem formation, a complete meristem will be formed, originating from a single cell and able to give rise to a green plant.[28]

We and others have been successful in establishing meristemic cultures of barley. It appears that the tissue culture technique is less genotype dependent though there are some varieties of which the cultures cannot be initiated. The first transformation experiments for barley using this method have just been reported. More studies should reveal whether indeed less somaclonal variation occurs than with the other methods.

2.4   Delivery of DNA

As the preview in the previous paragraphs already revealed, there are several methods for delivering the new genetic information (the DNA) to the target cells. Some methods work only in combination with a certain tissue. For instance, one cannot use particle bombardment on fragile protoplasts. The protoplasts, which have their cell wall removed, would not survive. It is also inconceivable that electroporation-mediated DNA transfer would work on cells with their cells walls still present. The DNA would merely get stuck in the cell

Fig. 2.4 Cultured meristematic tissue with shoots emerging from the periphery.

wall. This section will briefly describe which DNA transfer methods have been successfully used to transform barley and wheat.

2.4.1 Polyethyleneglycol (PEG) and electroporation

PEG and electroporation-mediated DNA transfer can be used only in combination with protoplasts derived from the different tissues described above. In an electroporator protoplasts are basically subjected to electric shock in a cuvette. The DNA is, due to this electric field, taken up by the protoplasts and some of this DNA is subsequently integrated into the genome of the protoplast. The settings of the electroporator are crucial to obtain some DNA uptake while minimising any damage to the protoplasts which will have to regenerate their cell wall to become a green plant. It is easy to find somewhat more rigorous settings to get DNA transfer but these will damage the protoplasts and their capacity to regenerate. Usually a whole set of conditions is tested with a reporter-DNA construct. Transfer of the reporter-DNA construct to the protoplasts means that these protoplasts are now capable of expressing an enzyme not normally present and of changing a colourless substrate to a blue product. By means of this process one can quickly determine which conditions transfer the DNA construct to the protoplast. One would then select the most gentle conditions that manage to transfer the DNA, assuming that these conditions would keep the damage to the protoplasts to a minimum.

Polyethyleneglycol-mediated DNA transfer works through a mechanism not completely understood, but it is thought that precipitation of the DNA on the plasma membrane of the protoplasts by calcium and PEG results in uptake of the DNA by the protoplasts. Large amounts of DNA are usually used to force uptake by the protoplasts. The PEG and calcium are diluted and washed away after 20–30 minutes. The whole procedure is quite rough on the protoplasts, as one can easily monitor using light microscopy. The protoplasts start out round but become quickly misformed in the PEG and calcium solution. It could take up to 24 hours after the removal of the PEG and calcium for them to become round and healthy again. The protoplasts will then regenerate their cell wall and start to divide.

2.4.2   Biolistics

The delivery of DNA to plant cells by means of biolistic methods has allowed the use of whole tissues as targets for transformation and therefore the first successful transformation procedures for wheat and barley. The principle of biolistics is very simple:

1. Gold or tungsten particles smaller than the plant cells are coated with DNA.
2. The target tissue is bombarded with the DNA-coated particles under vacuum.
3. The DNA diffuses from the particles in the plant cells and subsequently integrates into their genome.

Variation in this procedure can be: the type of particles, the coating procedure, and the speed of the particles when they hit the target tissue.[9,27,29] Many different biolistic procedures have been developed over time, the first one being literally a derivative of a gun where the particles were accelerated by gunpowder. Hence the name gene gun. The most recent gene guns control the speed of the particles by helium pressure.

One can imagine that particles with a high velocity can easily damage plant cells, and conditions have to be empirically determined for different target tissues. It has been shown that the particles usually damage the first cell layers of the target tissue. The particles stop in the underlying cells where they have apparently penetrated the cell wall but have not damaged the cells to a great extent. These cells might take up the DNA released by the particles in their genome. The next step is of course to select for the cell with the new genetic information as described in the next section. Target tissues used for particle bombardment-mediated transformation have so far been microspores, callus tissue, immature embryos and apical meristems.

2.4.3   *Agrobacterium*

Virulent strains of the soil bacterium *Agrobacterium tumefaciens* are capable of infecting a wide range of dicotyledonous plants and trees, which results in crown gall disease. The virulence capability of *Agrobacterium* is determined by the presence of extra-chromosomal genetic information contained on a plasmid, an autonomous circular piece of DNA. The plasmids of the various *Agrobacterium* strains have somewhat different characteristics. These plasmids have been manipulated by recombinant DNA techniques in such a way that new genetic information inserted into the plasmid will be transferred by the *Agrobacterium* to the infected plant cell without causing crown gall disease. It has been precisely determined which part of the *Agrobacterium* plasmid is transferred to the plant cells. In other words, the *Agrobacterium* can now be used as a carrier for genetic information to be introduced into the plant cell. The bacteria are then simply killed by antibiotics which do not affect the plant cell.

While it was originally thought that *Agrobacterium* was only capable of infecting dicotyledonous plants, it has recently become clear that some supervirulent strains of *Agrobacterium* are also capable of infecting wounded cells of monocots such as barley and wheat under laboratory conditions. In particular, immature embryo cells which have been wounded by gold particles from the gene gun are susceptible to *Agrobacterium* infection.[30,31] The highest transformation frequencies for barley are now obtained with *Agrobacterium*.

2.5   Selection and regeneration

2.5.1   Selectable markers

Delivery of DNA to target tissue results, in all cases, in a mixture of cells that are transformed and not transformed. It is therefore essential to select for the transformed cells and against the non-transformed cells. For wheat and barley this has been achieved by co-transforming with DNA encoding selectable markers. These selectable markers produce enzymes normally absent in wheat and barley that make the transformed cells resistant to either antibiotics or herbicides. Co-transformation means that DNA encoding the selectable markers is presented to the target tissue at the same time as the new genetic information to be introduced. The choice of the selectable marker and the corresponding selection agent is very limited for both wheat and barley. The selectable markers either confer resistance to antibiotics or herbicides.

1. *bar* gene from *Streptomyces* confers resistance to the herbicide Bialaphos.
2. *hpt* gene from *E.coli* confers resistance to the antibiotic hygromycin.
3. *nptII/aphA* gene from *E.coli* confers resistance to the antibiotics kanamycin, geneticin, G418 and paronomycin.
4. cp4/gox genes from bacterial origin confers resistance to the non-selective herbicide Roundup.

Before applying an antibiotic or herbicide to transformed target tissue, an applicable dose has to be determined on non-transformed tissue. One must be sure that non-transformed cells do not grow without overdosing the selective agent which could result in no growth of the transformed cells or loss of regeneration capacity. The right dose of selective agent has to be determined empirically for each target tissue. All four selectable markers have been successfully used to select for wheat transformants[32–34] and except for cp4/gox the other selectable markers have been used successfully for barley.[35] Each research group seems to have its own preference with regards to the selectable marker.

One of the requirements of a good selection system is that the selective agent can be applied to any cell type. This requires that the selectable markers are expressed in every cell type. The four selectable marker genes in the transformed plants have all been controlled by constitutive promoters. A promoter is a piece of genetic information that controls the expression of the neighbouring gene. A constitutive promoter is active in every cell type thus fulfilling the requirement set earlier. The activity of the promoter (see Section 2.6.1) determines how effectively the selectable marker is expressed.

2.5.2   Regeneration under selective conditions

The regeneration of the transformed cells to fully fertile plants is crucial to the transformation protocol. Before regeneration can take place, the transformed target tissue is maintained for a substantial time on selective medium which also contains auxins. The combination of auxins and selective agents favours the proliferation of transformed cells. The regeneration capacity of the transformed cells decreases usually with the period of tissue culture and the doses of the selectable agent and this has therefore to be minimised. The maintenance medium for barley has empirically been optimised by including for instance cytokinins.

Regeneration takes place on a medium without auxins but including cytokinin. Shoots develop on the proliferating cells which can then be transferred to a small pot with another synthetic medium in which roots develop. The regenerants will by then look like small cereal plants and can be transferred to soil. Regeneration is done in the absence or presence of the selective agent, depending on the tissue used. Some groups prefer a low selective pressure resulting in a large number of regenerants of which a high percentage are not transformed. Others prefer a stronger dose of the selective agent to minimise the number of regenerated plants that are not transformed. However, a high dose of the selective agent might affect the quality of the transformants in a negative way. In cases where tissue is used with a very high regenerative capacity such as apical meristem cultures or microspores, it is a prerequisite to have a strong effective selection before regeneration is induced. The selectable marker would therefore have to be controlled by a strong constitutive promoter. Figure 2.5 shows an apical meristem culture from which a small barley plant has started to regenerate.

Fig. 2.5 Meristem culture from which a small barley plant has started to regenerate.

2.6   Promoters

The genetic information that drives the expression of the selectable marker and the genetic information to be introduced are called the promoters. The promoter basically determines in which cells the introduced gene is expressed. As discussed above, the selectable marker should preferably be expressed in every cell to facilitate selection for transformed cells. Expression of the gene of interest may be required to be limited to a certain tissue such as the aleurone and endosperm. Promoters have been isolated that confer tissue-specific expression.

2.6.1   Constitutive promoters

The most commonly used constitutive promoters in wheat and barley transformation to drive either the selectable marker or a reporter gene are:[36]

- the 35S promoter from the Cauliflower mosaic virus (35S)
- the Actin 1 promoter from rice (Act1)
- the Ubiquitin 1 promoter from maize (Ubi1).

These promoters have different activities in wheat and barley. The 35S promoter has low activity though new enhanced versions are now available. The Act1 promoter has a moderate activity while the Ubi1 promoter has high activity in wheat and barley. The Ubi1 promoter would therefore be the promoter of choice for driving a selectable marker but the Act1 and 35S promoter might suffice as well.

Constitutively expressed promoters are of course also useful for expressing genes that confer pathogen resistance which has to be expressed throughout the plant. They might also find application in the overexpression of biomolecules with commercial value. Moreover they might be used for the anti-sense technique which can be used specifically to repress endogenous genes. In the anti-sense situation, the endogenous gene is constitutively expressed in the opposite, anti-sense orientation and therefore knocks out the endogenous (sense) gene, eliminating production of protein from this gene.

2.6.2   Tissue-specific promoters

Several wheat and barley promoters have been isolated that are likely to be expressed in specific tissues.[36] The effectiveness of these promoters, however, has not always been shown by transformation of wheat or barley. Some of these promoters have been analysed only in rice but it is very likely that these promoters function in wheat and barley in a similar fashion.

For instance, several α-amylase promoters have been isolated from wheat and barley. These promoters should in principle be expressed only in aleurone cells and the epithelium of the scutellum. The original data regarding the specificity of these promoters do not come from transformants but from a whole range of very convincing molecular biological data. It has recently indeed been

Table 2.1   Examples of seed-specific promoters

Tissue	Promoter	Source	Reference
Endosperm	Hordein	Barley	46, 47
	Glutenin	Wheat	48
Aleurone	Amylase	Barley, Wheat	49
	Glucanase	Barley	50, 51
	LTP1	Barley	52
Embryo	Lipoxygenase	Barley	53
Developing aleurone	LTP2	Barley	54

confirmed in transgenic rice that one of the α-amylases of rice is expressed in the epithelium scutellum early during germination and subsequently in the aleurone cells.

Table 2.1 shows a list of some seed-specific promoters that are available for wheat and barley. Using these promoters, expression of newly introduced genetic information can be limited to the tissue in which the promoter is active. In the developing seed expression can be limited to, e.g., the starchy endosperm. For both wheat and barley this means processes that take place in the endosperm, such as starch and protein synthesis, can in principle be influenced via a transgenic approach. This requires of course that genes encoding starch biosynthetic enzymes and storage proteins are available, which is indeed so. Figure 2.6 shows an example of a transgenic barley seed which has a reporter gene driven by a hordein promoter in its genome.[37] The reporter gene is able to change a colourless substrate to a blue product and, as depicted, this occurs only in the starchy endosperm (dark stain) and not in the embryo, scutellum, testa, pericarp or husk.

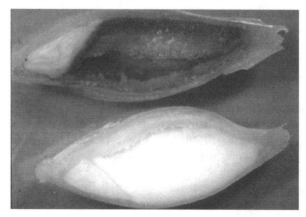

Fig. 2.6   Histochemical enzyme assay for β-glucuronidase (Gus) on non-transformed barley seeds and seeds from barley transformed with a Hordein-B1-Gus construct. Expression of Gus is only detected in the transformed seeds (top) in the endosperm (dark). Courtesy of M-J Cho, Department of Plant and Microbial Biology, University of California, Berkeley, USA.

2.7   Examples of transformed wheat and barley

2.7.1   Disease resistance

The first experiments to engineer disease resistance in barley focused on barley yellow dwarf virus. Wan and Lemaux (1994)[26] transformed barley with a construct containing the coat protein of the virus under control of the constitutive 35S promoter. This approach was based on the results of virus protection experiments with dicots. The experiments with the dicots showed that overexpression of the viral coat protein could result in viral protection. The mechanism of this protection is not completely understood though it is thought to act via silencing of the activity of the viral genome. Several of the transgenic barley lines were resistant to the barley yellow dwarf virus. However, no field trials have been conducted yet.

2.7.2   Malting related

Since barley is used for malting purposes to serve the brewing and distilling industry, a lot of effort has gone into transforming barley with malting-related genes. Barley has been transformed with a heat-stable 1,3-1,4-β-glucanase hybrid from *Bacillus*,[38] a heat-stable β-glucanase from the fungus *Trichoderma reesei*[39] and with mutagenised barley β-amylase[40] with higher heat stability. The corresponding endogenous barley enzymes are heat labile and their activities are destroyed either during kilning or mashing. In the case of 1,3-1,4-β-glucanase this might result in an extract with a high glucan content which is prone to give a beer with a haze.

The nucleotide sequence of the hybrid *Bacillus* β-glucanase was extensively modified, without altering the amino acid sequence, so that the codon usage was more like the endogenous β-glucanase. This was necessary since some codons are very rarely used in barley aleurone and would therefore limit expression levels. One of the α-amylase promoters was used to drive the expression of the heat-stable β-glucanase. The result was that the transformant indeed produced a heat-stable β-glucanase during germination that was absent in the control plants. The effect of the heat-stable β-glucanase on malting, however, has to be determined.

The fungal β-glucanase was used in unmodified form and driven by the α-amylase promoter as well. The protein was expressed and functional during germination though molecular weight and iso-electric points were different from the protein isolated from the fungus. The enzyme was thermotolerant which was revealed by β-glucanase assays at 65°C. The transgenic barley has not been used in malting studies yet but the enzyme has been exogenously applied during mashing and has proven to keep soluble glucans low and to improve filterability of the wort.

The mutagenised barley β-amylase with higher heat stability was driven by the endogenous β-amylase promoter. Several transformants were obtained that expressed a heat-stable β-amylase which was absent in the untransformed plants. The effect on malting and how the heat-stable β-amylase would alter the sugar spectra of worts remains to be investigated.

2.7.3   Nutritional quality

Barley grains have low contents of lysine and threonine and have therefore poor nutritional value for animals. The biochemistry of the biosynthetic pathways for threonine and lysine are well understood and it appears that there are two major regulatory enzymes: aspartate kinase (AK) and dihydrodipicolinate synthase (DHPS). Both enzymes in barley are subject to feedback inhibition by the end products, lysine and threonine. Mutant barley varieties have been identified in which AK lacks feedback inhibition but these barley mutants have not found any application for commercialisation. Brinch-Pedersen et al. (1996)[41] have taken a transgenic approach and have expressed feedback-insensitive AK and DHPS from E.coli in barley. Both E.coli genes were under control of the constitutive 35S promoter. Analysis of the transgenic plants showed that the leaves of the transgenic barley lines contained a fourteenfold increase in free lysine and eightfold increase in free methionine. Moreover, there was a twofold increase in lysine, arginine and asparagine in the mature seeds while free proline was reduced by 50%. No differences were observed in the composition of total free amino acids in the seeds. These results suggest that this transgenic barley would be of higher nutritional value than the non-transformed. The next step would be to introduce these genes into a current malting variety and to determine what effect the transgenes have on malting quality. If there is no effect on the malting quality then this barley would be of particular interest to farmers. It quite often occurs that the malting barley variety in the field is not really up to malting quality. This barley will be used for feeding animals and the farmers might get a better price for high-lysine/threonine barley.

2.7.4   Baking quality

Considerable amounts of fundamental research into the function of glutenins in relation to dough properties have accumulated. The processing characteristics of wheat dough are thought to be closely related to the number of active high-molecular-weight glutenin genes. High gluten doughs are in general more elastic and more suitable for making bread. The function and formation of the glutenin polymer and how these could be targeted by genetic engineering are reviewed by Vasil and Anderson.[42] Several groups have now taken transformation approaches to introduce extra glutenins into wheat. Blechl and Anderson[43] and Altpeter et al.[44] introduced various constructs containing high-molecular-weight glutenin subunits (HMW-GS) into the wheat variety Bobwhite. The expression of the HMW-GS under control of a glutenin promoter was clearly demonstrated but no data on the elasticity of the dough were presented. More recently, Barro et al.[45] showed that indeed dough elasticity increased with an increase in copies of HMW-GS. They transformed a wheat line containing less endogenous HMW-GS copies and the challenge now is to transform current cultivars which are already selected for bread-making quality to see whether the dough elasticity in these cultivars can also be improved.

2.8 Summary: problems and future trends

2.8.1 Technical difficulties

Besides the optimisation of the transformation protocols and the establishing of protocols for current elite varieties, there are three prominent problems when transformants are obtained:

1. somaclonal variation
2. agronomic performance
3. transgene stability and expression.

The term somaclonal variation is used for the appearance of phenotypes different from the transformed variety and unrelated to the transgene. This somaclonal variation is, as mentioned earlier, likely to be due to prolonged tissue culture in the presence of synthetic auxins and to selective agents. Moreover, tissue culture can also lead to a reduced agronomic performance of the regenerants (which do not necessarily have to be transformed). One would like to produce as many transgenic lines as feasible and pick the ones that have lost little or none of their agronomic performance and still show the phenotype from the transgene. Alternatively one could backcross the transformant and inbreed that line for a few generations while following the presence of the transgene at the molecular level and the agronomic performance of the crosses.

The stability of the transgene is also not guaranteed, neither is the expression. It has occurred that the transgene is lost in later generations. Moreover, it can also occur that the transgene is silenced, i.e. that expression of the transgene is repressed. So in spite of the presence of the transgene, no protein encoded by the transgene is produced in the transformants. It is therefore important to produce as many independent transformants as possible. Each independent transformant will have the transgene at a different position in its genome. One would then have to select for the lines that show stable expression of the transgene.

2.8.2 Acceptance of transgenic wheat and barley

The public perception is, according to the tabloids and broadsheets, not in favour of growing transgenic crops in the UK. Due to the pressure of the media a three-year moratorium on the commercial growing of transgenic crops has been announced by the Government. Moreover, food and drinks containing detectable transgenic ingredients have to be labelled as such. Some supermarket chains have announced that they will pull all GM foods off their shelves. Without debating the pros and cons of genetically modified cereals here, we would like to touch upon a few items concerning transgenic wheat and barley, i.e. the farmers, the industrial users and the consumers. After all, it is clear that import of transgenic cereals is not restricted.

Farmers may accept transgenic wheat and barley when economic advantages are offered. These economic advantages could either be in crop, pest or weed management, yield or growing wheat with special characteristics for a particular industry. In the absence of any weedy relatives, the introduction of herbicide resistance into wheat and barley might only raise concerns by organic farmers. The farmers will have to be sure that their crops will be bought by the industry, while the public would want to know whether the transgenic crops are as safe as non-transgenic crops.

Some industrial users of wheat and barley are already confronted with the import of transgenic maize. These industries sometimes use a mixture of maize and wheat or barley in their process. For instance, some beers are brewed using maize as an adjunct while some grain distillers use maize as well. At present there is plenty of non-transgenic maize available for these industries. The supply of non-transgenic maize might even be continued, with some processing companies in the USA now working with non-transformed maize only. But it might come with extra costs to the industry and consequently to the consumer.

2.8.3   Future trends

Data regarding the number and types of transgenic wheat and barley are very limited since companies tend to keep their experiments confidential. Traits that receive attention for both wheat and barley are disease resistance. It is likely that further attempts will be taken to engineer fungal and viral resistance. For wheat, another emphasis will be in starch biosynthesis and dough characteristics, while for barley emphasis might be on malting characteristics and nutritional value. Moreover, both barley and wheat could be used for the production of proteins with high value.

On the technological front, improved transformation protocols for wheat and barley will appear. The introduction of transgenes will also be done more carefully, so that no selectable markers are present in the transformants. It might also be possible in the future to transform large DNA molecules which carry information of multiple genes or a complete trait.

2.9   Sources of further information and advice

Birch, R G, Plant transformation: Problems and strategies for practical application. *Annual Review Plant Physiology Plant Molecular Biology*, 1997 **48** 297–326.

Bommineni, V R and Jauhar, P P, An evaluation of target cells and tissues used in genetic transformation of cereals. *Maydica*, 1997 **42** 107–20.

Christou, P, Strategies for variety-independent genetic transformation of

important cereals, legumes and woody species utilizing particle bombard-
ment. *Euphytica*, 1995 **85**(1–3) 13–27.
Lemaux, P G, Cho, M-J, Zhang, S and Bregitzer, P, 'Transgenic cereals:
Hordeum vulgare L. (barley)', in *Molecular Improvement of Cereals Crops*,
I.K. Vasil, ed. 1999, Kluwer Academic Publishers pp. 255–316.
The UK Plant Bioinformatics Network
    http//synteny.nott.ac.uk/
Quantitative trait loci detection
    http://www.css.orst.edu/barley/nabgmp/QTLFIG.HTM
GRAINGENES
    http://wheat.pw.usda.gov/
Agriculture genome information system
    http://probe.nalusda.gov:8000/
North American barley genome mapping project
    http://www.css.orst.edu/barley/nabgmp/nabgmp.htm
Cooperative Extension Specialist for the University of California at Berkeley
    http://plantbio.berkeley.edu/~outreach/

2.10   References

1. VASIL, V, CASTILLO, A, FROMM, M and VASIL, I, Herbicide resistant fertile
    transgenic plants obtained by micro-projectile bombardment of regenerable
    embryogenic callus. *Bio/Technology*, 1992 **9** 743–7.
2. VASIL, V, SRIVASTAVA, V, CASTILLO, A M, FROMM, M E and VASIL, I K, Rapid
    production of transgenic wheat plants by direct bombardment of cultured
    immature embryos. *Bio/Technology*, 1993 **11**(12) 1553–8.
3. WEEKS, J T, ANDERSEN, O D and BLECHL, A E, Rapid production of multiple
    independent lines of fertile transgenic wheat (*Triticum aestivum* L.). *Plant
    Physiology*, 1993 **102**(4) 1077–84.
4. RITALA, A, ASPEGREN, K, KURTÉN, U, SALMENKALLIO-MARTTILA, M, MANNO-
    NEN, L, HANNUS, R, KAUPPINEN, V, TEERI, T H and ENARI, T-M, Fertile
    transgenic barley by particle bombardment of immature embryos. *Plant
    Molecular Biology*, 1994 **24** 317–25.
5. JÄHNE, A, BECKER, D, BRETTSCHNEIDER, R and LÖRZ, H, Regeneration of
    transgenic, microspore derived, fertile barley. *Theoretical and Applied
    Genetics*, 1994 **89** 525–33.
6. NEHRA, N S, CHIBBAR, R N, LEUNG, N, CASWELL, K, MALLARD, C, STEINHAUER,
    L, BAGA, M and KARTHA, K K, Self-fertile transgenic wheat plants
    regenerated from isolated scutellar tissues following microprojectile
    bombardment with two distinct gene constructs. *Plant Journal*, 1994 **5**(2)
    285–97.
7. BARCELO, P, *et al.*, Transformation of wheat: State of the technology and
    examples of application. *9th International Wheat Genetics Symposium*,
    Saskatoon, Saskatchewan, Canada, University Extension Press, University

of Saskatchewan, 1998.

8.   CHO, M-J, JIANG, W and LEMAUX, P G, Transformation of recalcitrant barley cultivars through improvement of regenerability and decreased albinism. *Plant Science*, 1998 **138**(2) 229–44.

9.   KOPREK, T, HAENSCH, R, NERLICH, A, MENDEL, R R and SCHULTZE, J, Fertile transgenic barley of different cultivars obtained by adjustment of bombardment conditions to tissue response. *Plant Science*, 1996 **119**(1–2) 79–91.

10. BREGITZER, P, HALBERT, S E and LEMAUX, P G, Somaclonal variation in the progeny of transgenic barley. *Theoretical and Applied Genetics*, 1998 **96**(3–4) 421–5.

11. JIANG, W, CHO, M-J and LEMAUX, P G, Improved callus quality and prolonged regenerability in model recalcitrant barley (*Hordeum vulgare* L.) cultivars. *Plant Biotechnology*, 1998 **15**(2) 63–9.

12. LÜHRS, R and LÖRZ, H, Initiation of morphogenic cell-suspension and protoplast cultures of barley (*Hordeum vulgare* L.). *Planta*, 1988 **175** 71–81.

13. LAZZERI, P A, BRETTSCHNEIDER, R, LÜHRS, R and LÖRZ, H, Stable transformation of barley via PEG-induced direct DNA uptake into protoplasts. *Theoretical and Applied Genetics*, 1991 **81** 437–44.

14. KIHARA, M and FUNATSUKI, H, Fertile plant regeneration from barley (*Hordeum vulgare* L.) protoplasts isolated from primary calluses. *Plant Science*, 1995 **106** 115–20.

15. KIHARA, M and FUNATSUKI, H, Fertile plant regeneration from barley (*Hordeum vulgare* L.) protoplasts isolated from long-term suspension culture. *Breeding Science*, 1994 **44** 157–60.

16. FUNATSUKI, H, KURODA, H, KIHARA, M, LAZZERI, P A, MÜLLER, E, LÖRZ, H and KISHINAMI, I, Fertile transgenic barley generated by direct DNA transfer to protoplasts. *Theoretical and Applied Genetics*, 1995 **91** 707–12.

17. SALMENKALLIO-MARTTILA, M, ASPEGREN, K, ÅKERMAN, S, KURTÉN, U, MANNONEN, L, RITALA, A, TEERI, T H and KAUPPINEN, V, Transgenic barley (*Hordeum vulgare* L.) by electroporation of protoplasts. *Plant Cell Reports*, 1995 **15** 301–4.

18. NOBRE, J, DAVEY, M R and LAZZERI P A, Barley scutellum protoplasts: Isolation, culture and plant regeneration. *Physiologia Plantarum*, 1996 **98**(4) 868–74.

19. ZHANG, J, TIWARI, V K, GOLDS, T J, BLACKHALL, N W, COCKING, E C, MULLIGAN, B J, POWER, J B and DAVEY, M R, Parameters influencing transient and stable transformation of barley (*Hordeum vulgare* L.) protoplasts. *Plant Cell, Tissue and Organ Culture*, 1995 **441** 125–38.

20. FUNATSUKI, H and KIHARA, M, Influence of primary callus induction conditions on the establishment of barley cell suspensions yielding regenerable protoplasts. *Plant Cell Reports*, 1994 **13** 551–5.

21. HOEKSTRA, S, ZIJDERVELD VAN, M, BERGEN VAN, S, MARK VAN DER, F and HEIDEKAMP, F, Genetic modification of barley for end use quality.

Improvement of cereal quality by genetic engineering; Guthrie Centenary Conference of the Royal Australian Chemical Institute Cereal Chemistry Division, Sydney, New South Wales, Australia, Plenum Press, New York, 1994.

22. TOURAEV, A, VICENTE, O and HEBERLE-BORS, E, Initiation of microspore embryogenesis by stress. *Trends in Plant Science*, 1997 **2**(8) 297–302.

23. YAO, Q A, SIMION, E, WILLIAM, M, KROCHKO, J and KASHA, K J, Biolistic transformation of haploid isolated microspores of barley (*Hordeum vulgare* L.). *Genome*, 1997 **40** 570–81.

24. WAN, Y and LEMAUX, P G, 'Biolistic transformation of microspore derived and immature zygotic embryos and regeneration of fertile transgenic barley plants', in *Gene Transfer to Plants*, I. Potrykus and G. Spangenberg, eds. 1995, Springer-Verlag: Berlin, Germany; New York, USA. pp. 139–46.

25. BECKER, D, BRETTSCHNEIDER, R and LÖRZ, H, Fertile transgenic wheat from microparticle bombardment of scutellar tissue. *Plant Journal*, 1994 **5**(2) 299–307.

26. WAN, Y and LEMAUX, P G, Generation of large numbers of independently transformed fertile barley plants. *Plant Physiology*, 1994 **104** 37–48.

27. BARCELO, P and LAZZERI, P A, 'Transformation of cereals by microprojectile bombardment of immature inflorescence and scutellum tissues', in *Methods in Molecular Biology*, H. Jones, ed. 1995, Humana Press Inc.: Totowa, New Jersey, USA. pp. 113–22.

28. ZHANG, S, CHO, M-J, KOPREK, T, YUN, R, BREGITZER, P and LEMAUX, P G, Genetic transformation of commercial cultivars of oat (*Avena sativa* L.) and barley (*Hordeum vulgare* L.) using *in vitro* shoot meristematic cultures derived from germinating seedlings. *Plant Cell Reports*, 1999 **18**(12) 959–66.

29. MENDEL, R R, MÜLLER, B, SCHULZE, J, KOLESNIKOV, V and ZELENIN, A, Delivery of foreign genes to intact barley cells by high-velocity microprojectiles. *Theoretical Applied Genetics*, 1989 **78** 31–4.

30. CHENG, M, FRY, J E, PANG, S, ZHOU, H, HIRONAKA, C M, DUNCAN, D R, CONNER, T W and WAN, Y, Genetic transformation of wheat mediated by *Agrobacterium tumefaciens*. *Plant Physiology*, 1997 **115** 971–80.

31. TINGAY, S, McELROY, D, KALLA, R, FIEG, S, WANG, M, THORNTON, S and BRETTEL, R, *Agrobacterium tumefaciens*-mediated barley transformation. *Plant Journal*, 1997 **11** 1369–76.

32. ORTIZ, J P A, REGGIARDO, M I, RAVIZZINI, R A, ALTABE, S G, CERVIGNI, G D L, SPITTELER, M A, MORATA, N M, ELIAS, F R and VALLEJOS, R H, Hygromycin resistance as an efficient selectable marker for wheat stable transformation. *Plant Cell Reports*, 1996 **15**(12) 877–81.

33. ZHOU, H, *et al.*, Glyphosate-tolerant CP4 and GOX genes as a selectable marker in wheat transformation. *Plant Cell Reports*, 1995 **15** 159–63.

34. WITRZENS, B, BRETTELL, R I S, MURRAY, F R, McELROY, D, LI, Z and DENNIS, E S, Comparison of three selectable marker genes for transformation of wheat by microprojectile bombardment. *Australian Journal of Physiology*, 1998

25 39–44.

35. HAGIO, T, HIRABAYASHI, T, MACHII, H and TOMOTSUNE, H, Production of fertile transgenic barley (*Hordeum vulgare* L.) plant using the hygromycin-resistance marker. *Plant Cell Reports*, 1995 **14** 329–34.

36. McELROY, D and BRETTELL, R I S, Foreign gene expression in transgenic cereals. *Trends in Biotechnology*, 1994 **12**(2) 62–8.

37. CHO, M-J, CHOI, H W, BUCHANAN, B B and LEMAUX, P G, Inheritance of tissue-specific expression of barley hordein promoter-*uidA* fusions in transgenic barley plants. *Theoretical Applied Genetics*, 1999 **98** 1253–62.

38. JENSEN, L G, OLSEN, O, KOPS, O, WOLF, N, THOMSEN, K K and VON WETTSTEIN, D, Transgenic barley expressing a protein-engineerd, thermostable (1,3-1,4)-β-glucanase during germination. *Proc. Natl. Acad. Sci. USA*, 1996 **93**(4) 3487–91.

39. MANNONEN, L, RITALA, A, NUUTILA, A M, KURTÉN, U, ASPEGREN, K, TEERI, T H, AIKASALO, R, TAMMISOLA, J and KAUPPINEN, V, Thermotolerant fungal glucanase in malting barley. *26th European Brewing Convention Congress*, Maastricht, Netherlands, Oxford University Press, Oxford, 1997.

40. KIHARA, M, OKADA, Y, KURODA, H, SAEKI K, YOSHIGI, N and ITO, K, Generation of fertile transgenic barley synthesizing thermostable β-amylase. *26th European Brewing Convention Congress*, Maastricht, Netherlands, Oxford University Press, Oxford, 1997.

41. BRINCH-PEDERSEN, H, GALILI, G, KNUDSEN, S and HOLM, P B, Engineering of the aspartate family biosynthetic pathway in barley (*Hordeum vulgare* L.) by transformation with heterologous genes encoding feedback-insensitive aspartate kinase and dihydrodipicolinate synthase. *Plant Molecular Biology*, 1996 **32**(4) 611–20.

42. VASIL, I K and ANDERSON, O D, Genetic engineering of wheat gluten. *Trends in Plant Science*, 1997 **2**(8) 292–7.

43. BLECHL, A E and ANDERSEN, O D, Expression of a novel high-molecular-weight glutenin subunit in transgenic wheat. *Nature Biotechnology*, 1996 **14**(7) 875–9.

44. ALTPETER, F, VASIL, V, SRIVASTAVA, V and VASIL, I K, Integration and expression of the high-molecular-weight glutenin subunit 1Ax1 gene into wheat. *Nature Biotechnology*, 1996 **14**(9) 1155–9.

45. BARRO, F, ROOKE, L, BÉKÉS, F, GRAS, P, TATHAM, A S, FIDO, R, LAZZERI, P A, SHEWRY, P R and BARCELO, P, Transformation of wheat with high molecular weight subunit genes results in improved functional properties. *Nature Biotechnology*, 1997 **15**(11) 1295–9.

46. MARRIS, C, GOAALOIS, P, COPLEY, J and KREIS, M, The 5' flanking region of a barley B-Hordein gene controls tissue and developmental specific CAT expression in Tobacco plants. *Plant Molecular Biology*, 1988 **10**(4) 359–66.

47. FORDE, B G, HEYWORTH, A, PYWELL, J and KREIS, M, Nucleotide-sequence of a B1-Hordein gene and the identification of possible upstream regulatory elements in endosperm storage protein genes from barley, wheat and

maize. *Nucleid Acid Research*, 1985 **13**(20) 7327–39.

48. COLTO, V, BARTELS, D, THOMPSON, R and FLAVELL, R, Molecular characterization of an active wheat LMW glutenin gene and its relation to the wheat and barley prolamin genes. *Molecular & General Genetics*, 1989 **216**(1) 81–90.

49. BETHKE, P C, SCHUURINK, R C and JONES, R L, Hormonal signalling in cereal aleurone. *Journal Experimental Biology*, 1997 **48**(312) 1337–56.

50. LITTS, J C, SIMMONS, C R, KARRER, E E, HUANG, N and RODRIGUEZ, R L, The isolation and characterisation of a barley 1,3-1,4-β-glucanase gene. *European Journal of Biochemistry*, 1990 **194**(3) 831–8.

51. WOLF, N, Structure of the genes encoding *Hordeum vulgare* (1-3,1-4)-β-glucanase isoenzymes I and II and functional analysis of their promoters in barley aleurone protoplasts. *Molecular and General Genetics*, 1992 **234** 33–42.

52. SKRIVER, K, LEAH, R, MULLER-URI, F, OLSEN, F-L and MUNDY, J, Structure and expression of the barley lipid transfer protein gene LTP1. *Plant Molecular Biology*, 1992 **18**(3) 585–9.

53. ROUSTER, R, VAN MECHELEN, J and CAMERON-MILLS, V, The untranslated leader sequence of barley Lipoxygenase 1 gene (Lox1) confers embryo-specific expression. *Plant Journal*, 1998 **15**(3) 435–40.

54. KALLA, R, SHIMAMOTO, K, POTTER, R, NIELSEN, P S, LINNESTAD, C and OLSEN, O-A, The promoter of the barley aleuron-specific gene encoding a putative 7kDA lipid transfer protein confers aleuron cell-specific expression in transgenic rice. *Plant Journal*, 1994 **6**(6) 849–60.

3

The genetic transformation of rice and maize

M. R. Davey, H. Ingram, K. Azhakanandam and J. B. Power,
University of Nottingham

3.1   Introduction

A continued increase in food production will be essential in order to sustain an increasing world population. This must be achieved by the development, for example, of higher yielding varieties with improved nutritional quality and tolerance to biotic and abiotic stresses.[1] Whilst conventional breeding will play a major role in increasing crop yield, it is clear that laboratory-based techniques, such as genetic transformation to introduce novel genes into crop plants, will be essential in complementing existing breeding technologies. Since cloned genes are introduced into target cultivars during transformation, this eliminates the requirement for repeated sexual back-crossing to remove undesirable co-transferred genes. Importantly, the range of agronomically useful genes that can be introduced into cereals, such as rice, by transformation is more extensive than the range of genes which could be introgressed, using other somatic cell techniques such as interspecific somatic hybridisation (involving protoplast fusion), from the genome pool of wild species.

A decade ago, cereals were considered recalcitrant to transformation[2] but since that time, their genetic transformation has become more efficient, as considerable research has been focused on these crops because of their agronomic status. Globally, rice and maize are two of the major cereals, with rice providing the staple diet of more than one-third of the world's population. Consequently, it is not surprising that these plants have become prime targets for genetic manipulation involving the introduction of foreign DNA.

Initially, the genetic transformation of cereals, such as rice and maize, relied upon the use of experimental systems centred mainly on the ability of protoplasts isolated from suitable tissues of a limited range of target cultivars to take up DNA and to produce callus from which fertile plants could be regenerated. More recently, major advances have been accomplished in the regeneration of fertile plants from a range of source tissues, providing an essential foundation for the generation of transgenic plants. Interestingly, the transformation of rice and maize has progressed more rapidly than that of other cereals, which reflects the progress made in the tissue culture and regeneration of rice and maize. Thus, for several cultivars, the transformation technology is now routine, although procedures still need to be refined to maximise transformation of specific cultivars. The exploitation of transformation technology has already resulted in transgenic maize and rice plants expressing agronomically useful characteristics, such as resistance to herbicides, insects, fungi and viruses, examples of which are discussed later in this chapter (Section 3.5). The literature relating to maize and rice is now so extensive that it is impossible to include reference to all published work and experimental details in this chapter. Consequently, the citation of references is limited, where possible, to the most recent and relevant reports.

3.2   Approaches to the transformation of maize and rice

Birch[3] discussed the requirements for efficient transformation systems. Essentially, the transformation protocol must be cultivar-independent, technically simple and safe to operate. The target tissue must be regenerable, readily available and the time required in culture should be minimal to reduce both cost and somaclonal variation. Ideally, selection procedures must be efficient, producing stable transformants with simple integration patterns of foreign DNA at low copy number without incorporating unwanted vector sequences from outside the T-DNA, or encountering variable transgene expression, due to the position of insertion of the transgene or multiple copy-induced transgene co-suppression. The production of uniform transformants with efficient co-transformation and stable integration of multiple genes is essential. Additionally, the possibility of removing reporter and selectable marker genes from transformed plants is desirable and may become mandatory in the future.

Several experimental procedures are available to introduce genes into target plants, each approach having its own merits and limitations. Those methodologies which have been evaluated extensively for rice and maize, include DNA uptake into isolated protoplasts, the use of biolistics, electroporation, electrophoresis and silicon carbide whiskers and, more recently, *Agrobacterium*-mediated gene delivery.

3.2.1   Protoplast-based technologies

During the late 1980s and early 1990s, emphasis was placed on the use of protoplasts isolated from embryogenic cell suspensions of cereals as recipients for foreign DNA. Such procedures have been reviewed by Maas *et al.*[4] In general, embryogenic callus initiated from tissues, such as immature zygotic embryos, is transferred to a liquid medium for the production of rapidly growing, homogeneous cell suspensions containing a high proportion of densely cytoplasmic, actively dividing cells. Enzymatic digestion of the walls of such cells releases populations of protoplasts (naked cells), which are mixed with foreign DNA. Protoplasts take up this DNA through their plasma membrane when treated with chemicals such as polyethylene glycol (PEG) and/or high-voltage electrical pulses. Several difficulties are associated with this technique. Although about 50% of the protoplasts take up exogenous DNA, the final transformation frequency is low, and is usually in the order of 1 protoplast in 10^5 being stably transformed. Furthermore, the development of this technology relies upon efficient protoplast-to-plant regeneration systems, which are often genotype dependent. The production of cell suspensions which release toti-potent protoplasts is, theoretically, simple. However, in practice, it is laborious, time consuming, and relies heavily upon the intuition of the worker to identify and to select at an early stage of culture those cells which are most likely to produce cell suspensions suitable for protoplast isolation. In addition, aberrant plants are often regenerated from transformed protoplast-derived tissues,[5] frequently with multiple copies of the foreign DNA associated with complex patterns of integration into the plant genome.

Despite these limitations, protoplast transformation enabled production of the first transgenic rice[6–8] and maize plants.[9] Unfortunately, the transgenic plants regenerated in these early experiments were sterile. Subsequently, however, Golovkin et al.[10] introduced a mutant dihydrofolate reductase (DHFR) gene from mouse that confers methotrexate resistance into maize protoplasts using PEG and generated fertile plants, with cross-pollination experiments permitting segregation analysis of the transgene in seed generations. Interestingly, more recent studies have produced evidence that protoplast transformation can be used in maize to generate transformants with single-copy well-defined inserts of foreign DNA.[11] The latter authors introduced the virD1, virD2 and virE2 genes from the Ti plasmid of Agrobacterium tumefaciens into protoplasts on a plasmid, which carried the genes of interest flanked by the Agrobacterium 25 bp T-DNA repeat sequences. The latter were also from the Ti plasmid. The presence of the virE2 gene gave maximum transformation. Undoubtedly, such experiments were stimulated by recent advances in Agrobacterium-mediated transformation of cereals (see Section 3.2.3) and by improved knowledge of the molecular biology of Agrobacterium-plant interactions, especially the relevance of virulence (vir) genes in T-DNA transfer. In other experiments, Tsugawa et al.[12] reported the use of a synthetic polycationic amino polymer (polycation) for rice protoplast transformation, but this system requires further assessment. An extensive review of maize transformation using protoplasts is presented by Armstrong.[13] Similarly, Tyagi et al.[14] have summarised recent progress to 1997 in protoplast-mediated rice transformation for the production of herbicide-, fungal- and insect-resistant plants. It is likely that there could be a resurgence of interest in DNA uptake into isolated protoplasts for rice and maize transformation in view of the most recent achievements.

3.2.2   Biolistics

Biolistics, coined from the term 'biological ballistics', is also known as microprojectile bombardment, particle acceleration, particle bombardment or gene gunning. It involves the delivery of tungsten or gold particles (usually gold because they are chemically inert) of a suitable size ($0.4–1.2\mu m$) coated with DNA into plant cells. Since initial reports of this technology[15,16] the equipment has been improved to exploit electrical discharge[17] or helium pressure[18] for more controlled and consistent particle acceleration. Fertile, transgenic maize was first generated by Gordon-Kamm et al.[19] following microprojectile bombardment of embryogenic cell suspensions, while Christou et al.[20] initially applied this procedure to rice to introduce DNA into immature embryos of indica and japonica cultivars. Currently, this technology is in routine use for cereal transformation in several laboratories world-wide, utilising toti-potent cells as targets for transformation. As in the case of protoplast-mediated transformation, the literature reporting stably transformed rice and maize has been reviewed by Tyagi et al.[14] and Armstrong[13] respectively.

The advantages of microprojectile bombardment are that there is no requirement for the development of labour-intensive, cultivar-specific proto-plast-to-plant systems, and it bypasses any host specificity associated with Agrobacterium-mediated gene delivery. The system is both cultivar- and species-independent, simple to perform and transgenes can be introduced into any tissues of any plant genotype. The commercial availability of biolistic equipment, such as the BioRad PDS 1000/He device, has facilitated the standardisation of gene delivery parameters in different laboratories[21,22] although other devices, such as the particle inflow gun[23] or custom-built instruments, have been used to deliver genes into rice, maize and other cereals. Pareddy et al.[24] discussed several aspects of maize transformation by microprojectile bombardment using 'helium blasting' and described the principle and design of two novel devices for rapid DNA delivery and/or aiming capabilities in the production of transgenic maize. Noteworthy is the fact that Sudhakar et al.[25] have described the use of a portable, inexpensive helium-driven particle bombardment device, which is an improvement of the particle inflow concept and which operates without vacuum. Such an instrument, that gives transformation rates comparable to those of other more sophisticated devices, should facilitate the transfer and development of this technology to laboratories which, to date, have been unable to purchase more expensive instruments. Additionally, it may facilitate gene delivery to field-grown plants.

Important features relating to optimisation of transformation by biolistics include the nature of the microprojectiles, the use of calcium chloride and spermidine to aid adherence of DNA to microprojectiles, the choice of DNA construct and target explants, together with the physical bombardment parameters. The latter include the flight distance of particles in the instrument, the helium pressure and the vacuum conditions. The condition of the target cells is crucial in maximising transformation. For example, in the case of immature zygotic embryos, the donor plant growth conditions must be optimal, avoiding exposure to stress conditions such as pest or disease attack, or sub-optimal watering, temperatures and humidity. Poor donor plant growth conditions affect explant physiology and the competence of cells for transformation. The physical parameters and conditions influencing microprojectile bombardment are described by Christou[17,26] and Southgate *et al.*[22] Chen *et al.*[27] have reported their protocol for consistent, large-scale production of transgenic rice plants using biolistics, while Hagio[28] has provided suggestions for improving and maximising transformation efficiency using this gene delivery technique.

3.2.3 *Agrobacterium*-mediated transformation

The molecular biology of *Agrobacterium*-mediated transformation of dicotyledons is now relatively well understood[29] together with the possible mode of insertion of the DNA transferred from the Ti or Ri plasmid (the T-DNA) into the recipient plant genome.[30] An essential feature of the process is that the *virD1* and *virD2* genes of the Ti or Ri plasmid encode for endonucleases specific to the defined ends (25 bp borders) of the T-DNA, which ensure controlled DNA transfer with minimal rearrangement on entry into the recipient plant genome. Foreign DNA inserted between the borders of the T-DNA is introduced into recipient plant cells. Gene delivery into dicotyledons by *Agrobacterium tumefaciens*, and to a lesser extent by *A. rhizogenes*, has been exploited extensively for several years with T-DNAs, disarmed by removal of their oncogenicity genes, being used to effect gene delivery. Such disarmed T-DNAs are either retained on the Ti plasmid, or removed from the Ti plasmid and inserted into a smaller plasmid. The Ti plasmid, whilst deleted of its T-DNA, still retains its *vir* genes to effect T-DNA transfer from the smaller plasmid to recipient plant cells. Such Agrobacteria thus carry two plasmids, constituting the binary vector system. *Agrobacterium*-mediated gene delivery is simple, efficient when optimised, and transfers low copy number, intact transgenes.[31] However, the procedure has been more difficult to apply to cereals and other monocotyledons, since they are not infected naturally by *Agrobacterium*.

Although *Agrobacterium*-based procedures have been evaluated in several laboratories over a number of years, with early reports of T-DNA transfer into maize[32] and rice,[33] these early reports for *Agrobacterium*-mediated transformation of monocotyledons were viewed with scepticism. In a milestone publication, Hiei *et al.*[34] provided unequivocal evidence for the production of hygromycin-resistant GUS-expressing plants of japonica rice cultivars, which stimulated the development of *Agrobacterium*-mediated transformation systems for other cereals, such as maize. Undoubtedly, several factors influence the efficiency of *Agrobacterium*-mediated transformation of rice and maize. Specifically, the use of actively-dividing cells such as those in embryogenic scutellum-derived callus, exposure of bacterial cells to the phenolic wound signal molecule, acetosyringone, to activate *Agrobacterium vir* gene expression, the use of 'super-virulent' *Agrobacterium* strains and the development of 'super-binary' vectors[34,35] combined with an appropriate selectable marker.

In their report, Hiei *et al.*[34] evaluated a range of explants for transformation, together with *Agrobacterium* strains commonly used to transform
dicotyledons, namely LBA4404 and the 'super-virulent' EHA101. Transformation was most efficient with scutellum-derived embryogenic callus (in
common with the general observations for biolistics) of the japonica rice
cultivar Koshihikari. Both EHA101, carrying the binary vector pIG121Hm,
and LBA4404, with the vector pTOK233, transformed rice cells. LBA4404(p-
TOK233) contained, on the binary vector, the *virB* and *virG* genes from the *A.
tumefaciens* Ti plasmid Bo542, resulting in a 'super-binary' plasmid. This
strain/vector combination was superior to other strains evaluated, due to the
increased virulence resulting from the amplified expression of the *vir* genes.
Up to 28.6% of scutellum-derived tissues inoculated with LBA4404(p-
TOK233) produced transgenic plants. Subsequently, other workers have
reported reproducible *Agrobacterium*-mediated transformation of the indica
rice cultivars Basmati 385 and Basmati 370 by EHA101(pIG121Hm),[36] the
japonica cultivar Radon and the indica cultivars IR72 and TCS10 by
LBA4404(pTOK233),[37] and the commercially important USA javanica
cultivars Gulfmont and Jefferson[38] with the same constructs and strains
developed by Hiei *et al.*[34] Zhang *et al.*[39] also used LBA4404(pTOK233) to
generate fertile transgenic plants of the indica cultivar Pusa Basmati 1, while
Azhakanandam *et al.*[40] confirmed the efficiency of the same construct to
transform cultivars of japonica, indica and javanica sub-species of rice. All
these reports confirmed transgene integration into the genome of recipient
plants, generally at low copy number, although single copy T-DNA inserts
were rare.[40] Sequence analysis revealed that the borders of the T-DNA in
transgenic rice were essentially identical to those in *Agrobacterium*-
transformed dicotyledons.[34,38] Transmission of transgenes through seed
generations was Mendelian,[34] as confirmed by the other reports. Collectively,
these results confirmed the suitability of this experimental approach for the
biotechnological improvement of rice.[38]

Importantly, *Agrobacterium*-mediated transformation has been extended to
other cereals, with the first report of maize transformation by Ishida *et al.*[35]
using immature embryos as target tissues. An interesting concept is that of Trick
and Finer,[41] who showed that exposing tissues to brief periods of ultrasonication
in the presence of *Agrobacterium* results in a 100- to 1400-fold increase in
transgene expression in tissues of several crop plants, including maize.
Apparently, ultrasonication produces small and uniform fissures and channels
between plant cells, facilitating access of *Agrobacterium* into the internal plant
tissues. A similar approach has been reported by Zhang *et al.*[42] resulting in the
production of fertile maize plants transformed with an insecticidal protein (*Bt*)
gene from *Bacillus thuringiensis*, the acoustic intensity and duration of
treatment being important parameters in the procedure. Recent developments
in *Agrobacterium*-mediated cell transformation, including the production and
analysis of various *Agrobacterium* strains and vectors, are discussed by Tyagi *et
al.*,[14] Hiei *et al.*[43] and Komari *et al.*[44]

An interesting concept in vector development has been the use of procedures that combine the advantage of *Agrobacterium* transformation with the efficiency of biolistic delivery to cereals. Thus, Hansen and Chilton[45] described a novel 'agrolistic' system in which the *virD1* and *virD2* genes from *A. tumefaciens,* that are required in the bacterium for excision of T-strands from the T-DNA of the Ti plasmid during gene transfer to plant cells, were placed under the control of the CaMV35S promoter. The *vir* genes were co-delivered into maize cells by bombardment with a target plasmid containing T-DNA border sequences flanking the gene of interest. *Vir* gene products *in planta* caused strand-specific nicking at the right T-DNA border sequence, similar to *virD1/virD2* catalysed T-strand excision which normally occurs in *Agrobacterium*. Some inserts in transformed cells exhibited right border T-DNA junctions with plant DNA that corresponded precisely to the sequence expected for normal *Agrobacterium* T-DNA insertion events into the genome of dicotyledonous plant cells. It will be interesting to assess the applicability of this novel system for rice transformation.

3.2.4   Microinjection, silicon carbide whiskers and electrical procedures

The cost of reliable microinjection equipment combined with the difficulty of penetrating cell walls, and the need to avoid rupturing the vacuole allowing toxic vacuolar contents to enter the cytoplasm, have discouraged workers from applying this labour-intensive procedure to the transformation of cereals.[46] In spite of these limitations, Leduc *et al.*[47] reported the transient expression of the beta-glucuronidase (*gus*) gene and anthocyanin reporter genes following microinjection of excised maize zygotic embryos. In general, the real application of this technology is in assessments of gene expression following DNA injection into individual cells, since it is not a realistic procedure with which to generate large populations of transgenic plants. An interesting use of microinjection has been the introduction of individual *Agrobacterium* cells into single meristematic cells of maize to effect transformation.[48] Although young, immature maize embryos appear to lack competence for transformation, this approach proved that maize meristematic cells are, in fact, competent for transformation by *Agrobacterium*, although the response is genotype dependent.

The vortexing of embryogenic cell suspensions with DNA and silicon carbide fibres (whiskers), has been suggested as a simple, inexpensive procedure for the generation of transgenic plants.[49,50] Using this approach, Nagatani *et al.*[51] obtained 533 transformants in rice per gram fresh weight of target cells. Whilst other workers have assessed this procedure, they have obtained evidence for only very low levels of transgene expression in treated cells of maize and have been unable to regenerate transgenic plants.[22] The attraction of such a simple procedure is that it would be easy to transfer to other laboratories. However, evidence is yet to be presented for the reproducibility of this procedure.

Although the cell wall is often considered a barrier to DNA transfer, which may be overcome by removal or weakening of the wall by exposure to cellulytic and pectolytic enzymes, or by wounding, there are claims that electroporation (electropulsation) with high-voltage electrical pulses can deliver DNA to intact cells and tissues. In maize, this approach has been exploited to introduce the *gus* gene and genes for chloramphenicol acetyltransferase (*cat*) and phosphinothricin acetyltransferase (*bar*) into intact cells of black Mexican sweet maize,[52] cell plasmolysis before electropulsation being essential in the transformation process. Other workers have also transferred genes into immature embryos and Type II (embryogenic) callus of the inbred maize line A188.[22] The results of the latter authors were comparable to those of Pescitelli and Sukhapinda[53] for transgene (*gus*) expression in Type II callus of the 'High-II' hybrid and the 'back-crossed B73' genotypes. However, in contrast to the report of D'Halluin *et al.*[54] enzyme treatment of maize tissues was not required to facilitate DNA entry into recipient cells. Immersion of target cells in the electroporation buffer may be sufficient to permit diffusion of DNA through cell walls in readiness for its passage through the plasma membrane during electroporation.[53,22] Success has also been reported in transforming the elite Italian rice cultivars Lido, Cornaroli and Thaibonnet by electroporation of suspension cultured cells,[55] the transgenic plants exhibiting, importantly, negligible genomic changes.

Tissue electrophoresis has also been assessed for DNA delivery into cereals, DNA being believed to migrate into target tissues when the latter are exposed to a low voltage electrical field for several hours. Although there are reports of this procedure being used to introduce DNA into maize embryos, Southgate *et al.*[22] were unable to obtain evidence for gene introduction into embryogenic maize tissues. Laser beam-mediated gene transfer to rice has also been described.[56] Overall, whilst techniques such as microinjection, the use of silicon carbide whiskers, tissue electrophoresis and laser beams are available, they will probably continue to have limited application to cereal transformation.

3.3   Target tissues for rice and maize transformation

Successful plant transformation necessitates efficient DNA delivery into recipient cells, which must be capable of rapid cell division and plant regeneration. Consequently, the choice of tissue for transformation is crucial. Transformation of protoplasts by electroporation or PEG treatment requires a population of homogeneous cells for enzyme treatment and the release of protoplasts. Cell suspensions, leaf mesophyll tissue and aleurone layer cells have been used as a source of protoplasts, but, in general, embryogenic cell suspensions are most suitable for protoplast-based cereal transformation.[14] A long-term, non-regenerable suspension of the maize cultivar Black Mexican Sweetcorn (BMS) has been particularly useful in developing transformation procedures, in optimising protocols and in assessments of DNA constructs.[13] As already indicated, a major problem with protoplast transformation has been the production of sterile or morphologically abnormal plants from transformed tissues, which may be related to the long period (often several months) from initiation of the cultures to the time when the suspensions are homogeneous and the cells are in the correct developmental stage to release toti-potent protoplasts. In this respect, embryogenic scutellum-derived tissues may be most suitable for the rapid generation of cultures for protoplast production.[43]

Initially, immature embryos were used to develop procedures for micro-projectile bombardment of maize.[13] Klein *et al.*[16] conducted short-term (transient) gene expression studies with the maize cultivar A188, but failed to achieve stable transgene integration. Gordon-Kamm *et al.*[19] generated the first fertile, transgenic maize plants following microprojectile bombardment of embryogenic cell suspensions. Later, bombardment of immature embryo scutella[57] or callus[58] resulted in the production of fertile transgenic maize plants. Currently, immature embryos, immature embryo-derived callus and embryogenic cell suspensions are the main targets for microprojectile bombardment-mediated transformation of maize,[13] although Zhong *et al.*[59] reported a novel and reproducible transformation system using cultured shoot apices. Additional factors that influence transformation include explant pre-culture to initiate cell division prior to transformation, and/or treatment with a high osmoticum medium to induce cell plasmolysis. The latter reduces cell damage and leakage of cytoplasm on impact of the high-velocity DNA-carrying particles.[2,22] As Songstad *et al.*[60] stated, the advantage of bombarding scutellum-derived tissues is their ability to regenerate shoots rapidly, the latter exhibiting increased male and female fertility compared to shoots regenerated from long-term callus or cell suspensions.

The earliest report of a cultivar-independent method for biolistic-mediated transformation of rice was by Christou *et al.*[20] who targeted immature zygotic embryos. Subsequently, Cao *et al.*[61] bombarded embryogenic cell suspensions. As in the case of maize, the targets currently used for routine transformation of rice include immature embryos, embryogenic callus and embryogenic cell suspensions, derived from immature or mature embryos.[44] The review by Tyagi *et al.*[14] gives further details describing the biolistic-mediated transformation of rice.

In evaluating several tissues of rice for *Agrobacterium*-mediated transformation, Hiei *et al.*[34] concluded that scutellum-derived embryogenic calli gave the highest transformation frequency. Suspension cultures were unsuitable for *Agrobacterium* transformation unless the cells were transferred from liquid to semi-solid culture medium for a period prior to bombardment. Later, Aldemita and Hodges[37] described the use of immature zygotic embryos for efficient *Agrobacterium*-mediated transformation, while Park *et al.*[62] inoculated isolated shoot apices of rice to produce transgenic plants. The review papers of Hiei *et al.*[43] and Tyagi *et al.*[14] give comprehensive details of reports describing *Agrobacterium*-mediated transformation of rice. In extending *Agrobacterium*-mediated transformation to maize, Ishida *et al.*[35] followed the procedures originally developed for rice, using co-cultivation of immature maize embryos with bacterial cells. Maize transformation using *Agrobacterium* is still in its early stages and most research has been limited to the cultivar A188.[13] Less commonly used target cells for cereal transformation include microspores, anther culture-derived tissues and leaf bases.[2,14]

3.4   Vectors for rice and maize transformation

The development of cereal transformation technology has relied on the use of a limited number of constructs composed of constitutive promoters, reporter genes and selectable marker genes. More recently, efforts have focused on the identification of tissue-specific and developmentally regulated promoters, in combination with agronomically useful genes. Vectors for basic studies of transformation and the development of reliable protocols usually consist of a reporter gene and a selectable marker gene, each driven by a suitable promoter. Subsequently, vectors have also included genes for agronomically useful characteristics.

Promoter sequences are essential in order to control gene expression in target plant cells. Since the constitutively expressed cauliflower mosaic virus (CaMV) 35S promoter has been used extensively and successfully in the transformation of dicotyledons, it follows that this promoter has also been evaluated for cereal transformation. As already discussed, the 'super-binary' vector pTOK233 has been particularly useful in rice transformation by *A. tumefaciens*,[34] the vector carrying the hygromycin phosphotransferase (*hpt*) and *gus* genes, both with the CaMV35S promoter. In general, the CaMV35S promoter, combined with the *gus* reporter gene and the *hpt* selectable marker gene, has been the focus of *Agrobacterium*- and protoplast-based transformation procedures, whereas a larger range of promoters have been utilised in microprojectile bombardment studies of both rice and maize. Overall, the 35S promoter has proved to be less effective in monocotyledons than in dicotyledons. Consequently, several alternative promoters have been tested for rice and maize transformation.[22,63] Those used initially were reported by Fromm *et al.*,[64] McElroy *et al.*,[65] Reggiardo *et al.*,[66] Christensen *et al.*[67] and Chamberlain *et al.*[68] and are summarised in Table 3.1.

The cereal-derived *act1*, *ubi1* and *emu* promoters are particularly effective, since naturally they drive gene expression to high levels in maize and rice. They have been assessed in studies of rice and/or maize transformation, in association with reporter and selectable marker genes, including the *gus*, *bar*, *hpt* and neomycin phosphotransferase (*npt*II) genes (Table 3.2).[44,69] Compared to the 35S promoter, the *ubi1*, *act1* and *emu* promoters gave improved transient (short-term) expression following microprojectile bombardment of rice[69] and maize.[70] The introduction of introns associated with promoter sequences in genes constructed for cereal transformation has enhanced expression in transformed tissues. For example, Vain *et al.*[71] compared a range of monocot and dicot intron-containing fragments inserted into the 5′ untranslated leader between the promoter and transgene (*gus*) of interest. In this study, the maize *ubi1* intron gave a 71-fold enhancement of gene expression in maize. Interestingly, enhancement of gene

Table 3.1 Promoters used in cereal transformation studies

Promoter	Construct	Reference
CaMV35S-*adh1 intron1*	CaMV35S promoter, enhanced with the first intron from the maize alcohol dehydrogenase (adh) gene	Fromm *et al.*[64]
act1-act1 intron1	Rice actin 1 promoter	McElroy*et al.*[65]
adh1-adh1 intron1	Maize alcohol dehydrogenase promoter	Reggiardo *et al.*[66]
ubi1-ubi1 intron1	Maize ubiquitin promoter	Christensen *et al.*[67]
emu	Modified *adh1* promoter and first intron	Chamberlain *et al.*[68]

Table 3.2 Selectable marker genes for cereal transformation

Gene	Enzyme encoded	Confers resistance to
*npt*II	Neomycin phosphotransferase	Neomycin, kanamycin, geneticin/G418, paromomycin
hpt	Hygromycin phosphotransferase	Hygromycin B
Bar	Phosphinothricin acetyltransferase	Phosphinothricin (PPT), bialaphos, glufosinate ammonium, Basta®

expression was not limited to introns from monocotyledons, since the dicot *chsA* intron gave a 92-fold enhancement of *gus* expression in maize. However, as these authors emphasised, it does not necessarily follow that an intron which is effective in maize will be equally effective in other cereals.

Since major advances have been accomplished in recent years in our knowledge of rice genetics,[14] this cereal is now considered a model for both *Agrobacterium*- and biolistics-mediated transformation of cereals. Consequently, research is more advanced in rice than in other monocotyledons. Current strategies are focused on the identification and development of tissue-specific promoters. Maize and rice transformation studies have involved several of these promoters, including tapetum-specific promoters (e.g. *Osg6B*) and endosperm-specific promoters (maize *waxy* gene, zein, *CM3*). Additional details of vector constructs for specific studies in rice and maize have been reviewed comprehensively by Bommineni and Jauhar,[2] Armstrong,[13] Tyagi *et al.*,[14] Hiei *et al.*,[43] Komari *et al.*[44] and McElroy and Brettell.[63]

The promoter most suitable for driving transgene expression can be confirmed by linking, at least initially, the promoter to a suitable reporter gene, followed by assessments of gene expression during short-term or longer periods (stable transformation). A reporter gene should have no negative effects on plant metabolism, its product should be stable *in vivo* and the detection assay should be simple, low cost and sensitive. Any endogenous gene activity should be minimal in order not to mask reporter gene expression. The *gus*, anthocyanin and luciferase (*luc*) reporter genes have been used in rice and maize, the most commonly used being the *gus* gene, derived from *Escherichia coli*. Beta-glucuronidase, the enzyme encoded by the *gus* gene, catalyses the hydrolysis of a range of substrates, including X-Gluc (5-bromo-4-chloro-3-indolyl-β-D-glucuronic acid), the product of which forms an indigo dye within transformed plant cells.[72] The disadvantage of this assay is that it is destructive. Nevertheless, the *gus* assay is an improvement on previous systems, such as the CAT assay, which requires radioactive reagents.[14] Other reporter genes used for assessments of both transient and stable transformation include the *npt*II gene whose expression can be assessed by ELISA, and the C1, B and R genes which regulate anthocyanin biosynthesis. Their expression results in red pigmentation in transformed cells. The luciferase gene from the firefly (*Photinus pyralis*), which catalyses the oxidation of D(-)-luciferin in the presence of ATP, leads to the accumulation of oxyluciferin and emission of yellow light, although visualisation requires expensive equipment.[63] A relatively new marker is the green fluorescent protein (gfp) from the jellyfish *Aequora victoria*, the assay of which is simple, requires no additional substrate and is non-destructive.[73-75]

Selectable marker genes permit the growth of only transformed cells on selective medium, usually by conferring resistance to antibiotics and herbicides. Those commonly used for rice and maize transformation are summarised in Table 3.2. The *npt*II gene, conferring resistance on plant cells to antibiotics such as kanamycin sulphate and geneticin (G418), was used in early studies of DNA uptake into cereal protoplasts and in *Agrobacterium*-mediated gene delivery. However, cells of cereals can be naturally resistant to kanamycin sulphate and this antibiotic also inhibits shoot regeneration from transformed tissues. Similarly, selection on a medium containing G418 results in poor organogenesis, although this antibiotic is less inhibitory than kanamycin to shoot regeneration. Those shoots which do regenerate in the presence of kanamycin or G418 are often albinos. The *hpt* gene, from *E. coli*, encoding for hygromycin resistance, is more effective for rice.[43] Additionally, the *bar* gene from *Streptomyces hygroscopicus* is both a selectable marker and introduces an agronomically useful characteristic, herbicide resistance, into transformed cells and regenerated plants.[14,44]

It is interesting to note that genomic DNA can also be used to effect transformation. Thus, Sawahel[76] used total genomic DNA isolated from a hygromycin-resistant maize cell line to transform hygromycin-sensitive lines to antibiotic resistance. The merit of this procedure is that it circumvents the requirement for gene cloning and, consequently, could see application in the future for the introduction of agronomically useful genes into target cultivars. A possible limitation may be the size of DNA used, since this appears to influence transformation.[77]

3.5   Examples of agronomically useful genes introduced into rice and maize

Understanding the expression of simple reporter and selectable marker genes in transgenic plants is important in predicting the behaviour of agronomically useful genes introduced into crops. Consequently, it is imperative to assess gene expression in large seed populations. In this respect, Zhong et al.[59] studied expression of the *bar*, potato proteinase inhibitor II and *uidA* (*gus*) genes, confirming their co-integration, co-inheritance and co-expression in 286 first generation (T1) plants and in 11,000 second seed generation (T2) maize plants. Similarly, Brettschneider et al.[21] followed the inheritance and expression of transgenes to the fourth seed generation in several inbred lines and sexual hybrids of maize. In field studies, Oard et al.[78] assessed expression of the *bar* gene, giving resistance to the herbicide glufosinate, in the commercial rice cultivars Gulfmont, IR72 and Koshihikari. They confirmed that the *bar* gene was effective in conferring field-level resistance to the herbicide in rice, although, importantly, variation amongst transgenic lines required traditional breeding selection procedures to identify plants with high levels of herbicide resistance. These workers also emphasised the need to generate several independent transgenic lines of each cultivar for transgene assessments.

Recent studies have reported the development of transgenic plants containing agronomically useful genes, in addition to those for herbicide resistance. Since insects cause substantial crop losses world-wide, it follows that engineering plants for insect resistance has, and will continue, to receive high priority. Stem-boring insects are common pests in maize and rice, and resistance against these insects has been achieved, primarily, by the introduction and expression of modified or synthetic versions of the *Bt* δ-endotoxin, a natural insecticidal toxin from *Bacillus thuringiensis*. For example, Wunn et al.[79] and Cheng et al.[80] introduced the *cryIA(b)* gene into rice cultivars, including the indica cultivar IR58, to confer resistance to yellow stem borer (*Scirpophaga incertulas*) and striped stem borer (*Chilo suppressalis*), while Alam et al.[81] were the first to engineer a lowland deep water rice for stem borer resistance using the same gene. In addition to giving resistance to stem-boring insects, expression of the *cryIA(b)* gene also inhibited feeding of the leaf-folding insects *Cnaphalo crocis* and *Marasmia patnalis* on transgenic rice.[79] Comparisons have been made of the expression of the *Bt cryIA(b)* gene driven by different promoters, including the constitutive 35S and Actin-1 promoters, with tissue-specific promoters from pith tissue and the pep-carboxylase (PEPC) promoter from chlorophyllous tissue of maize. The latter promoter gave high levels of transgene expression in leaves and stems of rice.[82]

Modified versions of the *cryIA(b)* gene have been used in rice transformation to give plants which induced 100% mortality in feeding yellow stem borers.[83] Synthetic truncated genes, based on the *cryIA(b)* gene, have also been introduced into rice.[84] The latter authors targeted low tillering aromatic rices which are particularly difficult to improve by conventional breeding because of loss of quality characteristics upon sexual hybridisation. The *cryIA(c)* gene has also been assessed in rice transformation for stem borer resistance;[80,85] the *cry2A Bt* gene also conferred resistance to yellow stem borer and rice leaf folder insects in the indica rices Basmati 370 and M7.[86] A recent example of the transformation of maize for insect resistance is that of Fearing *et al.*[87] who introduced the *cryIA(b)* gene into six commercial cultivars and four back-cross generations. They reported the highest concentration of insecticidal protein to be at anthesis in transformed plants. Other genes conferring insect resistance which have been evaluated in rice, include the Cowpea trypsin inhibitor (*CpTi*) gene, which increased resistance of transgenic rice to striped stem borer and the pink stem borer,[88] and the snowdrop lectin (GNA) gene. The latter was directed against sap-sucking insects, such as the brown plant hopper, through the use of a rice sucrose synthase promoter to drive GNA expression in the phloem of transgenic plants.[89]

Nematodes cause severe crop losses in some areas, including rice cultivated in Africa. In attempts to reduce nematode damage, a cysteine proteinase inhibitor (oryzacystatin-I Delta D86) gene was introduced into four elite African rice cultivars (ITA212, IDSA6, LAC23, WAB56-104), resulting in a 55% reduction in egg production by the nematode *Meloidogyne incognita* in the roots of transgenic plants.[90]

Viral and fungal diseases reduce crop productivity. The insertion of viral coat protein genes into transgenic plants is a well-established procedure for conferring viral resistance, this approach being exploited in rice for resistance to rice dwarf virus.[91] Rice has also been engineered for resistance to sheath blight incited by the fungus *Rhizoctonia solani*. Thus, introduction of a 1.1kb fragment of a rice chitinase gene linked to the CaMV35S promoter resulted in transgenic plants in which resistance to the fungus correlated directly with chitinase activity.[92] It will be interesting to determine whether chitinase gene expression confers cross-protection to other fungal pathogens. More recently, the introduction of the stilbene synthase gene, which is thought to be involved in the synthesis of a phytoalexin, provided protection in rice to infection by the fungus *Pyricularia oryzae*.[93]

One of the challenges facing biotechnologists is to modify plants so as to increase net carbon gain.[94] C_4 plants, such as maize and several weed species, have evolved a biochemical mechanism to overcome oxygen inhibition of photosynthesis. In an initial assessment of the feasibility of improving photosynthesis in C_3 plants, the intact gene of phospho*enol*pyruvate carboxylase (PEPC), which catalyses the initial fixation of atmospheric carbon dioxide in maize, was introduced into japonica cultivars of rice, a C_3 plant.[94] Transgenic plants exhibited reduced oxygen inhibition of photosynthesis and photosynthetic rates comparable to those of non-transformed plants. Such an approach for modifying one of the major physiological processes in plants holds promise for the transformation of cultivars of the other sub-groups (indica, javanica) of rice and, indeed, for the transformation of other C_3 crops. Other experiments have been reported which attempt to increase the resistance of crop plants to environmental stresses such as ozone, high light, drought, cold and heat. A common feature in stressed plants is the production of free oxygen radicals which damage DNA, lipids and proteins. In this respect, transformation procedures have been presented to increase the levels of superoxide dismutase, ascorbate peroxidase and catalase in cells in order to improve the tolerance of maize and rice to oxidative stress.[95]

Experiments have been directed towards modifying the nutritional quality of rice grain. For example, introduction of a fatty acid desaturase gene from tobacco into rice resulted in modification of the proportions of linoleic acid and linolenic acid in fatty acids, with the former decreasing and the latter increasing, respectively.[96] A significant recent advance has been the transformation of rice to produce beta-carotene, a precursor of vitamin A. Thus, the introduction of genes for phytoene synthase, phytoene desaturase, carotene desaturase and lycopene cyclase from *Narcissus,* or a double-desaturase from the fungus *Erwinia uredovora*, resulted in transgenic plants with grain producing yellow endosperm.[97] Some lines produced enough beta-carotene to supply the daily human requirements from 300 grams of cooked rice. Since vitamin A deficiency affects about 7% of the world population (mostly children), mainly in developing countries, this work represents a significant advance in attempts to alleviate the problem of vitamin A deficiency.

In the future, it may be essential to engineer complex biochemical pathways by the introduction of several transgenes into target species. In order to provide a foundation for this technology, embryogenic tissues of rice have been bombarded with a mixture of 14 different genes on pUC-based plasmids.[98] Eighty per cent of the regenerated plants contained more than two and 17% more than nine of the transgenes. Importantly, plants with transgenes were phenotypically normal and 63% set viable seed. Detailed information collected over several seed generations from such plants will clarify the interaction and expression of multiple transgenes in genetically engineered plants, such as cereals.

3.6   Summary: problems, limitations and future trends

Despite progressing from basic research involving the development of transformation protocols to the introduction and expression of useful genes in several rice and maize cultivars, cereal transformation technology still has its limitations. For example, technology transfer between laboratories and between individual workers is particularly difficult in cereals. Additionally, reduced expression or failure of transgenes to express, is reflected in low transformation frequencies.[3,17] Failure of transgenes to express may be related to positional effects associated with the integration of foreign DNA into the recipient plant genome[31,99] and to transgene silencing. Transgene silencing is caused by cytosine methylation, an inherent control mechanism for gene expression.[63] It is often associated with sense-suppression of multiple integrated copies of the transgene, or to co-suppression of an endogenous gene. In rice transformed by particle bombardment, Kohli *et al.*[100] presented evidence that truncation of transgenes produces incomplete transcripts giving aberrant RNA species, which may also be responsible for transgene silencing. Other workers[101] have also shown that transgene silencing in rice occurs at the transcriptional level, although no evidence was found for gross alterations or methylation of CCGG sites in a chitinase transgene driven by the CaMV35S promoter. Kumpatla and Hall[102] have stressed the need for detailed molecular analyses to be performed over several seed generations on plants harbouring transgenes and that lines containing methylated inserts should be carefully evaluated before being included in long-term breeding strategies. Plants transformed at a single locus should be more amenable to breeding programmes, as the single transgenic locus will be easier to characterise genetically.[103] These authors stated that particle bombardment generally results in a single transgenic locus, with this site acting as a 'hot spot' for subsequent integration of successive transforming DNA molecules.

Currently, there is no simple procedure for the rapid identification of transgenic plants. Indeed, detailed molecular analysis, usually involving Southern blotting procedures[104] of the DNA isolated from individual plants from large populations of regenerants, is still essential in order to identify individuals with single copy gene inserts. Transformation with vectors carrying scaffold attachment regions (SARs) or matrix attachment regions (MARs), chromatin boundary elements that insulate genes from the effects of the surrounding chromatin, should assist in reducing the variability of transgene expression in transgenic plants.[14] Other problems associated with the transformation of cereals include reduced access to new technologies, genes and promoters due to intellectual property and patents rights, which limit research progress.[3,17] It is clear that attention must be paid to the gene delivery technology to minimise genomic and/or phenotypic effects. In this respect, Arencibia *et al.*[55] performed detailed molecular analyses on rice transformed by particle bombardment and intact cell electroporation and concluded, importantly, that their transgenic plants exhibited negligible genomic changes as a result of the transformation process. They emphasised the need to select appropriate techniques to generate plants which express foreign genes of interest, whilst retaining existing agronomic and/or industrial traits.

Some of the recent commercial releases of transgenic plants are listed by Birch[3] and discussed by Dunwell.[105,106] Before such releases are possible, field analysis is compulsory to determine genetic stability, long-term effectiveness in the field and to assess any potential hazards associated with transgenic plants and their products. Presently, there is considerable public concern for the risks of genetic engineering, especially in Europe and the UK. Some of the potential problems associated with this technology have been summarised by Boulter[107] and Rogers and Parkes.[108] Such risks include the establishment of weedy, feral crop populations, transgene escape to wild species through sexual hybridisation to produce new and uncontrollable herbicide-resistant weeds, pests or diseases, and the production of novel allergenic or toxic compounds by transgenic plants.

The escape of transgenes by pollen dispersal and through sexual hybridisation can be prevented by plastid transformation, since these organelles are not transmitted, in most plants, through pollen. However, the success of this procedure will necessitate the development of systems in which gene constructs are targeted to plastid DNA.[109] These authors have also provided evidence that certain genes, such as *Bt* genes for insect resistance, are expressed to higher levels in plastids than following nuclear integration. It should not be overlooked that many of the potential risks relating to genetic engineering of crops, which are currently being discussed, can also be associated with traditionally bred cultivars. It is of interest that the latter have been in existence for long periods without any cause for concern.

Risks are also posed by the cultivation of a homogeneous crop containing a useful gene, such as the *Bt* gene. This will increase selection pressure generating *Bt* toxin-resistant insect biotypes, leading to the rapid development of insects which can tolerate the *Bt* toxin and which are, as a consequence, able to overcome the resistance of transgenic plants. Strategies to reduce this risk include the use of several versions of the *Bt* toxin gene, reducing selection pressure for the development of resistance to a single form of the toxin.[3] Companies generating herbicide-resistant crops emphasise that the use of transgenic plants will reduce normally damaging and polluting herbicide applications by encouraging the use of safer, biodegradable herbicides. However, the availability of plants with resistance to a specific herbicide could lead, conversely, to overuse of the herbicide. A reduction in the broad spectrum of herbicides currently employed in agricultural practice could stimulate the development of herbicide-resistant weeds, as a result of selection pressure. The use of antibiotic-resistance genes to select transformed cells prior to the regeneration of transgenic plants, also poses potential hazards of horizontal gene transfer,[110] including the risk of transfer of the antibiotic resistance genes to bacteria in the gastro-intestinal tract.

The production of marker-free transgenic plants is becoming an important consideration[43] as a consequence of the concerns associated with herbicide and antibiotic resistance genes. Yoder and Goldsbrough[99] detail the approaches available in marker-free technology. In summary, methods exist to excise a marker gene from transgenic plants, including the use of prokaryotic site-specific recombination systems, such as the bacteriophage P1 *Cre/lox* system, where a transgenic plant is generated containing two *lox* sequences flanking the selectable marker. The *Cre* gene, introduced through a second transformation, catalyses recombination between two *lox* sequences, resulting in excision of the undesirable marker gene. Transposable elements can also be used to excise marker genes in a similar manner. For example, the maize *Ac* transposable element encodes for transposase, which trans-activates the *Ds* transposable element causing its excision and reinsertion into a new locus. Any marker gene sequences cloned between the inverted repeats of a *Ds* element, are also activated. Studies have shown that 10% of excised elements do not reinsert, leading to loss of the marker gene.[99] Perhaps the most simple approach for generating marker-free transgenic plants is that of co-transformation.[43] Thus, transformation with two plasmids, with the undesirable marker gene on a separate plasmid from the useful gene, will allow selection of marker-free plants following segregation in seed progeny. The success of this process requires that the two transgenes are co-transformed efficiently and into unlinked sites. Finally, tissue- or temporal-specific expression will permit control of marker gene expression in transgenic plants.[14]

Current objectives for cereal transformation must be in response to the requirements of plant breeders who are directed by the product consumer. Such objectives include improvements in crop yield, pest and disease resistance, stress tolerance and nutritional quality. These improvements will necessitate the identification and cloning of agronomically useful genes prior to their introduction into specific crops.[13,14,17,111] Undoubtedly, one of the major research objectives must remain the manipulation of cereals to fix their own nitrogen[112] since this would have considerable environmental impact by reducing the application of nitrogenous fertilisers normally required to maximise crop yield. Improvements in vector design and construction are still required in order to generate transgenic plants in which single copies of the gene(s) of interest are inserted precisely into the plant genome and which express reliably throughout seed generations. Overall, transformation procedures should be simple, in order to facilitate technology transfer, and independent of plant genotype. Such transformation technology must be underpinned by reliable tissue culture procedures in order to identify those cells which are most competent for transformation and which are capable of regenerating into phenotypically normal, fertile transgenic plants. In addition, the time of culture should be minimal.[3] The nature and timing of the culture parameters are probably more important for the recovery of transgenic cells and plants than the selection procedure.

At present, particularly in the UK, consumer acceptance of genetically engineered crops, including cereals, poses one of the chief limitations to the exploitation of the technology discussed in this chapter. Clearly, the introduction of transgenic crops for human consumption must proceed slowly and with caution in order to accumulate reliable data on questions, such as the frequency of transgene introgression from transgenic crops to wild species, in order to build up public acceptance and confidence in the safety of such products. This should be combined with improved communication, by workers actively involved in gene manipulation technologies, of a balanced and unbiased view regarding transformation technology, in order to bridge any gaps in knowledge between scientists and the public.[107]

3.7   Sources of further information and advice

The references cited in the text are the prime source of information relating to the transformation of rice, maize and other cereals. The review articles are particularly helpful in this respect, since they provide detailed overviews of the subject. All published articles carry the address of the workers who have been responsible for conducting the experiments and collating the data. Most authors will provide detailed information, upon request, relating to aspects of their published procedures; many welcome visits to their laboratories to facilitate technology transfer at the practical level. In addition, it is now the policy of most international journals to include the e-mail address, fax and telephone numbers of communicating scientists in order to facilitate communication in this rapidly advancing field of research. Several internet sites are also available, which are useful in keeping workers informed of the current advances in the field of transgenic research in relation to crop improvement. For example, The Bowditch Group produces a News Bulletin to up-date researchers on the latest developments, particularly by agrochemical companies, in plant biotechnology (*http://www.bowditchgroup.com/index.html/feedback.htm*), while the University of Amsterdam also collates similar information (*http://www.pscw.uva.nl/monitor*). Other useful internet sites are given in the papers by Dunwell[105,106] and below.

3.8   Internet sites

Plant biotechnology websites: *www.agro.agri.umn.edu/plant-tc/optc.htm*

Organisations/Universities involved in agricultural research: *www.cgiar.org/isnar/arow/index.htm*

AgBiotechNet: *www.agbio.cabweb.org/home.htm*

International Association for Plant Tissue Culture: *www.hos.ufl.edu/ikvweb/iaptcb.htm*

Food safety – UK Government: *www.maff.gov.uk/food/foodindx.htm*

UK Patent Office: *www.patent.gov.uk*

US Department of Agriculture – biotechnology permits: *www.aphis.usda.gov/bbep/bpl*

US Patent Office: *www.uspto.gov/*

3.9   References

1. KHUSH G S, Origin, dispersal, cultivation and variation of rice, *Plant Mol Biol*, 1997 **35** 25–34.
2. BOMMINENI V R and JAUHAR P P, An evaluation of target cells and tissue used in genetic transformation of cereals, *Maydica*, 1997 **42** 107–20.
3. BIRCH R G, Plant transformation: problems and strategies for practical application, *Ann Rev Plant Physiol Plant Mol Biol*, 1997 **48** 297–326.
4. MAAS C, REICHEL C, SCHELL J and STEINBIß H-H, Preparation and transformation of monocot protoplasts, *Methods in Cell Biol*, 1995 **50** 383–99.

5. DAVEY M R, KOTHARI S L, ZHANG H, RECH E L, COCKING E C and LYNCH P T, Transgenic rice: characterisation of protoplast-derived plants and their seed progeny, *J Exp Bot*, 1991 **42** 1159–69.

6. TORIYAMA K, ARIMOTO Y, UCHIMIYA H and HINATA K, Transgenic plants after direct gene transfer into protoplasts. *Bio/Technol*, 1988 **6** 1072–84.

7. ZHANG W and WU R, Efficient regeneration of transgenic rice plants from rice protoplasts and correctly regulated expression of the foreign gene in the plants, *Theor Appl Genet*, 1988 **6** 835–40.

8. ZHANG H M, YANG H, RECH E L, GOLDS T J, DAVIS A S, MULLIGAN B J, COCKING E C and DAVEY M R, Transgenic rice plants produced by electroporation-mediated plasmid uptake into protoplasts, *Plant Cell Rep*, 1988 **7** 379–84.

9. RHODES C A, LOWE K S and RUBY K L, Plant regeneration from protoplasts isolated from embryogenic maize cell cultures, *Bio/Technol*, 1988 **6** 56–60.

10. GOLOVKIN M V, ABRAHAM M, MOROCZ S, BOTTKA S, FEHER A and DUDITS D, Production of transgenic maize plants by direct DNA uptake into embryogenic protoplasts, *Plant Sci*, 1993 **90** 41–52.

11. HANSEN G, SHILLITO R D and CHILTON M D, T-strand integration in maize protoplasts after codelivery of a T-DNA substrate and virulence genes, *Proc Natl Acad Sci USA*, 1997 **94** 11726–30.

12. TSUGAWA H, OTSUKI Y and SUZUKI M, Efficient transformation of rice protoplasts mediated by a synthetic polycationic amino polymer, *Theor Appl Genet*, 1998 **97** 1019–26.

13. ARMSTRONG C L, The first decade of maize transformation: a review and future perspective, *Maydica*, 1999 **44** 101–9.

14. TYAGI A K, MOHANTY A, BAJAJ S, CHAUDHURY A and MAHESHWARI S C, Transgenic rice: A valuable monocot system for crop improvement and gene research, *Critical Rev Biotechnol*, 1999 **19** 41–79.

15. SANFORD J C, KLEIN T M, WOLF E D and ALLEN N, Delivery of substances into cells and tissues using a particle bombardment process, *J Part Sci and Technol*, 1987 **5** 27–37.

16. KLEIN T M, WOLF E D, WU R and SANFORD J C, High-velocity microprojectiles for delivering nucleic acids into living cells, *Nature*, 1987 **327** 70–3.

17. CHRISTOU P, Strategies for variety-independent genetic transformation of important cereals, legumes and woody species utilizing particle bombardment, *Euphytica*, 1995 **85** 13–37.

18. SANFORD J C, DE VIT M J, RUSSELL J A, SMITH F D, HARPENDING P R, ROY M K and JOHNSTON S A, An improved helium-driven biolistic device, *Technique*, 1991 **3** 3–16.

19. GORDON-KAMM W J, SPENCER T M, MANGANO M L, ADAMS T R, DAINES R J, START W G, O'BRIEN J V, CHAMBERS S A, ADAMS JR. W R, WILLETTS N G, RICE T B, MACKEY C J, KRUEGER R W, KAUSCH A P and LEMAUX P G, Transformation of maize cells and regeneration of fertile transgenic plants, *Plant Cell*, 1990 **2** 603–18.

20. CHRISTOU P, FORD R and KOFRON M, Production of transgenic rice (*Oryza sativa* L.) plants from agronomically important indica and japonica

varieties via electric discharge particle acceleration of exogenous DNA into immature zygotic embryos, *Bio/Technol*, 1991 **9** 957–62.

21. BRETTSCHNEIDER R, BECKER D and LÖRZ H, Efficient transformation of scutellar tissue of immature maize embryos, *Theor Appl Genet*, 1997 **94** 737–48.

22. SOUTHGATE E M, DAVEY M R, POWER J B and WESTCOTT R J, A comparison of methods for direct gene transfer into maize (*Zea mays* L.), *In Vitro Cell Dev Biol – Plant*, 1998 **34** 218–24.

23. O'KENNEDY M M, BURGER J T and WATSON T G, Stable transformation of Hi-II maize using the particle inflow gun, *South Afr J Sci*, 1998 **94** 188–92.

24. PAREDDY D, PETOLINO J, SKOKUT T, HOPKINS N, MILLER M, WETTER M, SMITH K, CLAYTON D, PESCITELLI S and GOLD A, Maize transformation via helium blasting, *Maydica*, 1997 **42** 143–54.

25. SUDHAKAR D, DUC L T, BONG B B, TINJUANGJUN P, MAQBOOL S B, VALDEZ M, JEFFERSON R and CHRISTOU P, An efficient rice transformation system utilising mature seed-derived explants and a portable, inexpensive particle bombardment device, *Transgenic Res*, 1998 **7** 289–94.

26. CHRISTOU P, Rice transformation: bombardment, *Plant Mol Biol*, 1997 **35** 197–203.

27. CHEN L, ZHANG S, BEACHY R N and FAUQUET C M, A protocol for consistent, large scale production of fertile transgenic rice plants, *Plant Cell Rep*, 1998 **18** 25–31.

28. HAGIO T, Optimising the particle bombardment method for efficient genetic transformation, *Japan Agric Res Quart*, 1998 **32** 239–47.

29. SMITH R H and HOOD E E, *Agrobacterium tumefaciens* transformation of monocotyledons, *Crop Sci*, 1995 **35** 301–9.

30. TINLAND B, The integration of T-DNA into plant genomes, *Trends Plant Sci*, 1996 **1** 178–84.

31. GELVIN S B, The introduction and expression of transgenes in plants, *Curr Opin Biotechnol*, 1998 **9** 227–32.

32. SHEN W-H, ESCUDERO J, SCHLAPPI M, RAMOS C, HOHN B and KOUKOLIKOVA-NICOLA Z, T-DNA transfer to maize cells: Histochemical investigation of β-glucuronidase activity in maize tissues, *Proc Natl Acad Sci USA*, 1993 **90** 1488–92.

33. CHAN M-T, CHANG H-H, HO S-L, TONG W-F and YU S-M, *Agrobacterium*-mediated production of transgenic rice plants expressing a chimeric α-amylase promoter/β-glucuronidase gene, *Plant Mol Biol*, 1993 **22** 491–506.

34. HIEI Y, OHTA S, KOMARI T and KUMASHIRO T, Efficient transformation of rice (*Oryza sativa* L.) mediated by *Agrobacterium* and sequence analysis of the boundaries of the T-DNA, *Plant J*, 1994 **6** 271–82.

35. ISHIDA Y, SAITO H, OHTA S, HIEI Y, KOMARI T and KUMASHIRO T, High efficiency transformation of maize (*Zea mays* L.) mediated by *Agrobacterium tumefaciens*, *Nature Biotechnol*, 1996 **14** 745–50.

36. RASHID H, YOKOI S, TORIYAMA K and HINATA K, Transgenic plant

production mediated by *Agrobacterium* in indica rice, *Plant Cell Rep*, 1996 **15** 727–30.

37. ALDEMITA R R and HODGES T K, *Agrobacterium tumefaciens*-mediated transformation of japonica and indica rice varieties, *Planta*, 1996 **199** 612–17.

38. DONG J, TENG W, BUCHHOLZ W G and HALL T C, *Agrobacterium*-mediated transformation of Javanica rice, *Mol Breed*, 1996 **2** 267–76.

39. ZHANG J, XU R-J, BLACKLEY D, ELLIOTT M C and CHEN D-F, *Agrobacterium*-mediated transformation of elite indica and japonica rice cultivars, *Mol Biotechnol*, 1997 **8** 223–31.

40. AZHAKANANDAM K, McCABE M S, POWER J B, LOWE K C, COCKING E C and DAVEY M R, T-DNA transfer, integration, expression and inheritance in rice: effects of plant genotype and *Agrobacterium* super-virulence, *J. Plant Physiol*, 2000 (in press).

41. TRICK H N and FINER J J, SAAT: Sonication-assisted *Agrobacterium*-mediated transformation, *Transgenic Res*, 1997 **6** 329–36.

42. ZHANG H, WANG G Y, XIE Y, DAI J R, XU N, ZHAO N M, LI T Y, TIAN Y C, QIIAO L Y and MANY K Q, Transformation of maize embryogenic calli mediated by ultrasonication and regeneration of fertile transgenic plants, *Sci in China Series C – Life Sciences*, 1997 **40** 316–22.

43. HIEI Y, KOMARI T and KUBO T, Transformation of rice mediated by *Agrobacterium tumefaciens*, *Plant Mol Biol*, 1997 **35** 205–18.

44. KOMARI T, HIEI Y, ISHIDA Y, KUMASHIRO T and KUBO T, Advances in cereal gene transfer, *Curr Opin Plant Biol*, 1998 **1** 161–5.

45. HANSEN G and CHILTON M D, 'Agrolistic' transformation of plant cells: Integration of T-strands generated *in planta*, *Proc Natl Acad Sci USA*, 1996 **93** 14978–83.

46. SONGSTAD D D, SOMERS D A and GRIESBACH R J, Advances in DNA delivery techniques. *Plant Cell Tiss Org Cult*, 1995 **40** 1–15.

47. LEDUC N, MATTHYS ROCHON E, ROUGIER M, MOGENSEN L, HOLM P, MAGNARD J L and DUMAS C, Isolated maize zygotes mimic *in vivo* embryonic development and express microinjected genes when cultured *in vitro*, *Dev Biol*, 1996 **177** 190–203.

48. ESCUDERO J, NEUHAUS G, SCHLAPPI M and HOHN B, T-DNA transfer in meristematic cells of maize provided with intracellular *Agrobacterium*, *Plant J*, 1996 **10** 355–60.

49. THOMPSON J A, DRAYTON P R, FRAME W B, WANG K and DUNWELL J M, Maize transformation utilising silicon-carbide whiskers – A review, *Euphytica*, 1995 **85** 75–80.

50. WANG K, DRAYTON P, FRAME W B, DUNWELL J M and THOMPSON J, Whisker-mediated plant transformation – an alternative technology, *In Vitro Cell Dev Biol – Plant*, 1995 **31** 101–4.

51. NAGATANI N, HONDA H, SHIMADA T and KOBAYASHI T, DNA delivery into rice cells using silicon carbide whiskers, *Biotechnol Techniques*, 1997 **11** 471–3.

52. SABRI N, PELISSIER B and TEISSIE J, Transient and stable electrotransformations of intact black Mexican sweet maize cells are obtained after preplasmolysis, *Plant Cell Rep*, 1996 **15** 924–8.

53. PESCITELLI S M and SUKHAPINDA K, Stable transformation via electroporation in maize Type II callus and regeneration of fertile transgenic plants, *Plant Cell Rep*, 1995 **14** 712–16.

54. D'HALLUIN K, BONNE E, BOSSUT M, DE BEUCKELEER M and LEEMANS J, Transgenic maize plants by tissue electroporation, *Plant J*, 1992 **4** 1495–505.

55. ARENCIBIA A, GENTINETTA E, CUZZONI E, CASTIGLIONE S, KOHLI A, VAIN P, LEECH M, CHRISTOU P and SALA F, Molecular analysis of the genome of transgenic rice (*Oryza sativa* L.) plants produced via particle bombardment or intact cell electroporation, *Mol Breed*, 1998 **4** 99–109.

56. GUO Y D, LIANG H and BERNS M W, Laser-mediated gene-transfer in rice, *Physiol Plant*, 1995 **93** 19–24.

57. KOZIEL M G, BELAND G L, BOWMAN C, CAROZZI N B, CRENSHAW R, CROSSLAND L, DAWSON J, DESAI N, HILL M, KADWELL S, LAUNIS K, LEWIS K, MADDOX D, McPHERSON K, MEGHJI M R, MERLIN E, RHODES R, WARREN G W, WRIGHT M and EVOLA S V, Field performance of elite transgenic maize plants expressing an insecticidal protein derived from *Bacillus thuringiensis*, *Bio/Technol*, 1993 **11** 194–200.

58. WALTERS D A, VETSCH C S, POTTS D E and LUNDQUIST R C, Transformation and inheritance of a hygromycin phosphotransferase gene in maize plants, *Plant Mol Biol*, 1992 **18** 189–200.

59. ZHONG H, SUN B L, WARKENTIN D, ZHANG S B, WU R, WU T Y and STICKLEN M B, The competence of maize shoot meristems for integrative transformation and inherited expression of transgenes, *Plant Physiol*, 1996 **110** 1097–107.

60. SONGSTAD D D, ARMSTRONG C L, PETERSEN W L, HAIRSTON B and HINCHEE M A W, Production of transgenic maize plants and progeny by bombardment of Hi-II immature embryos, *In Vitro Cell Dev Biol – Plant*, 1996 **32** 179–83.

61. CAO J, DUAN X, McELROY D and WU R, Regeneration of herbicide resistant transgenic rice plants following microprojectile-mediated transformation of suspension culture cells, *Plant Cell Rep*, 1992 **11** 586–91.

62. PARK S H, KINSON S R M and SMITH R H, T-DNA integration into genomic DNA of rice following *Agrobacterium* inoculation of isolated shoot apices, *Plant Mol Biol*, 1996 **32** 1135–48.

63. McELROY D and BRETTEL R I S, Foreign gene expression in transgenic cereals, *Trends Biotechnol*, 1994 **12** 62–8.

64. FROMM M E, MORRISH F, ARMSTRONG C, WILLIAMS R, THOMAS J and KLEIN T M, Inheritance and expression of chimeric genes in the progeny of transgenic maize plants, *Bio/Technol*, 1990 **8** 833–9.

65. McELROY D, ZHANG W, CAO C and WU R, Isolation of an efficient *actin* promoter for use in rice transformation, *Plant Cell*, 1990 **2** 163–71.

66. REGGIARDO M I, ARANA J L, ORSARIA L M, PERMINGEAT H R, SPITTELER A and VALLEJOS R H, Transient transformation of maize tissues by micropro-jectile bombardment. *Plant Sci*, 1991 **75** 237–43.

67. CHRISTENSEN A H, SHARROCK R A and QUAIL P H, Maize polyubiquitin genes – structure, thermal perturbation of expression and transcript splicing, and promoter activity following transfer to protoplasts by electroporation, *Plant Mol Biol*, 1992 **18** 675–89.

68. CHAMBERLAIN D A, BRETTELL R I S, LAST D I, WITRZENS B, McELROY D, DOLFERUS R and DENNIS E R, The use of the *Emu* promoter with antibiotic and herbicide resistance genes for the selection of transgenic wheat callus and rice plants, *Aust J Plant Physiol*, 1994 **21** 95–112.

69. LI Z, UPADHYAYA N M, MEENA S, GIBBS A J and WATERHOUSE P M, Comparison of promoters and selectable marker genes for use in indica rice transformation, *Mol Breed*, 1997 **3** 1–4.

70. SCHLEDZEWSKI K and MENDEL R R, Quantitative transient gene expression: comparison of the promoters for maize polyubiquitin 1, rice *Actin* 1, maize derived *Emu* and CaMV 35S in cells of barley, maize and tobacco, *Transgenic Res*, 1994 **3** 249–55.

71. VAIN P, FINER K R, ENGLER D E, PRATT R C and FINER J J, Intron-mediated enhancement of gene expression in maize (*Zea mays* L.) and bluegrass (*Poa pratensis* L.), *Plant Cell Rep*, 1996 **15** 489–94.

72. JEFFERSON R A, KAVANAGH T A and BEVAN M W, GUS fusions: β-glucuronidase as a sensitive and versatile gene fusion maker in higher plants, *EMBO J*, 1987 **6** 3901–7.

73. CHALFIE M, TU Y, EUSKIRCHEN G, WARD W W and PRASHER D C, Green fluorescent protein as a marker for gene expression, *Science*, 1994 **263** 802–5.

74. VAIN P, WORLAND B, KOHLI A, SNAPE J W and CHRISTOU P, The green fluorescent protein (GFP) as a vital screenable marker in rice transformation, *Theor Appl Genet*, 1998 **96** 164–9.

75. SCHENK P M, ELLIOTT A R and MANNERS J M, Assessment of transient gene expression in plant tissues using the green fluorescent protein as a reference, *Plant Mol Biol Rep*, 1998 **16** 313–22.

76. SAWAHEL W, Molecular analysis of genomic DNA-mediated transformation in *Zea mays*, *Biol Plant*, 1997 **39** 361–7.

77. FLEMING G H, KRAMER C M, LE T and SHILLITO R D, Effect of DNA fragment size on transformation frequencies in tobacco (*Nicotiana tabacum*) and maize (*Zea mays*), *Plant Sci*, 1995 **110** 187–92.

78. OARD J H, LINSCOMBE S D, BRAVERMAN M P, JODARI F, BLOUIN D C, LEECH M, KOHLI A, VAIN P, COOLEY J C and CHRISTOU P, Development, field evaluation, and agronomic performance of transgenic herbicide resistant rice, *Mol Breed*, 1996 **2** 359–68.

79. WUNN J, KLOTI A, BURKHARDT P K, GHOSH-BISWAS G C, LAUNIS K, IGLESIAS V A and POTRYKUS I, Transgenic indica rice breeding line IR58 expressing a synthetic *cryIA(b)* gene from *Bacillus thuringiensis* provides effective

insect control, *Bio/Technol*, 1996 **14** 171–6.

80. CHENG X, SARDANA R, KAPLAN H and ALTOSAAR I, *Agrobacterium*-transformed rice plants expressing synthetic *cryIA(b)* and *cryIA(c)* genes are highly toxic to striped stem borer and yellow stem borer, *Proc Natl Acad Sci USA*, 1998 **95** 2767–72.

81. ALAM M F, DATTA K, ABRIGO E, VASQUEZ A, SENADHIRA D and DATTA S K, Production of transgenic deepwater indica rice plants expressing a synthetic *Bacillus thuringiensis cryIA(b)* gene with enhanced resistance to yellow stem borer, *Plant Sci*, 1998 **135** 25–30.

82. DATTA K, VASQUEZ A, TU J, TORRIZO L, ALAM MF, OLIVA N, ABRIGO E, KHUSH G S and DATTA S K, Constitutive and tissue-specific differential expression of the *cryIA(b)* gene in transgenic rice plants conferring resistance to rice insect pest, *Theor Appl Genet*, 1998 **97** 20–30.

83. WU C, FAN Y, ZHANG C, OLIVA N and DATTA S K, Transgenic fertile japonica rice plants expressing a modified *cryIA(b)* gene resistant to yellow stem borer, *Plant Cell Rep*, 1997 **17** 129–32.

84. GHAREYAZIE B, ALINIA F, CORAZON A, MENGUITO C A, RUBIA L G, DE PALMA J M, LIWANAG E A, COHEN M B, KHUSH G S and BENNETT J, Enhanced resistance to two stem borers in an aromatic rice containing a synthetic *cryIA(b)* gene, *Mol Breed*, 1997 **3** 401–14.

85. NAYAK P, BASU D, DAS S, BASU A, GHOSH D, RAMAKRISHNAN N A, GHOSH M and SEN S K, Transgenic elite indica rice plants expressing *CryIAc* delta-endotoxin of *Bacillus thuringiensis* are resistant against yellow stem borer (*Scirpophaga incertulas*), *Proc Natl Acad Sci USA*, 1997 **94** 2111–16.

86. MAQBOOL S B, HUSNAIN T, RIAZUDDIN S, MASSON L and CHRISTOU P, Effective control of yellow stem borer and rice leaf folder in transgenic rice indica varieties Basmati 370 and M7 using the novel delta-endotoxin *cry2A Bacillus thuringiensis* gene, *Mol Breed*, 1998 **4** 501–7.

87. FEARING P L, BROWN D, VLACHOS D, MEGHJI M and PRIVALLE L, Quantitative analysis of *CryIA(b)* expression in Bt maize plants, tissues, and silage and stability of expression over successive generations, *Mol Breed*, 1997 **3** 169–76.

88. XU D, XUE Q, McELROY D, MAWAL Y, HILDER V A and WU R, Constitutive expression of a cowpea trypsin inhibitor gene, *CpTi*, in transgenic rice plants confers resistance to two major rice insect pests, *Mol Breed*, 1996 **2** 167–73.

89. SUDHAKAR D, FU X D, STOGER E, WILLIAMS S, SPENCE J, BROWN D P, BHARATH M, GATEHOUSE J A and CHRISTOU P, Expression and immunolocalisation of the snowdrop lectin, GNA, in transgenic rice plants, *Transgenic Res*, 1998 **7** 371–8.

90. VAIN P, WORLAND B, CLARKE M C, RICHARD G, BEAVIS M, LIU H, KOHLI A, LEECH M, SNAPE J, CHRISTOU P and ATKINSON H, Expression of an engineered cysteine proteinase inhibitor (Oryzacystatin-I Delta D86) for nematode resistance in transgenic rice plants, *Theor Appl Genet*, 1998 **96** 266–71.

91. ZHENG H H, LI Y, YU Z H, LI W, CHEN M Y, MING X T, CASPER R and CHEN Z L, Recovery of transgenic rice plants expressing the rice dwarf virus outer coat protein (S8), *Theor Appl Genet*, 1997 **94** 522–7.

92. LIN W, ANURATHA C S, DATTA K, POTRYKUS I, MUTHUKRISHNAN S and DATTA S K, Genetic-engineering of rice for resistance to sheath blight, *Bio/Technol*, 1995 **13** 686–91.

93. STARK-LORENZEN P, NELKE B, HÄNSSLER G, MÜHLBACH H P and TOMZIK J E, Transfer of a grapevine stilbene synthase gene to rice (*Oryza sativa* L.), *Plant Cell Rep*, 1997 **16** 668–73.

94. KU M S B, AGARIE S, NOMURA M, FUKAYAMA H, TSUCHIDA H, ONO K, HIROSE S, TOKI S, MIYAO M and MATSUOKA M, High-level expression of maize phospho*enol*pyruvate carboxylase in transgenic rice plants, *Nature Biotechnol*, 1999 **17** 76–80.

95. VAN BREUSEGEM F, VAN MONTAGU M and INZE D, Engineering stress tolerance in maize, *Outlook on Agric*, 1998 **27** 115–24.

96. WAKITA Y, OTANI M, IBA I C and SHIMADA T, Co-segregation of an unlinked selectable marker gene and NtFAD3 gene in transgenic rice plants produced by particle bombardment, *Genes and Genet Systems*, 1998 **73** 219–26.

97. BURKHARDT P K, BEYER P, WÜNN J, KLÖTI A, ARMSTRONG G A, SCHLEDZ M, LINTIG J V and POTRYKUS I, Transgenic rice (*Oryza sativa*) endosperm expressing daffodil (*Narcissus pseudonarcissus*) phytoene synthase accumulates phytoene, a key intermediate of provitamin A biosynthesis, *Plant J*, 1997 **11** 1071–8.

98. CHEN L, MARMEY P, TAYLOR N J, BRIZARD J-P, ESPINOSA C, D'CRUZ P, HUET H, ZHANG S, DE KOCHKO A, BEACHY R N and FAUQUET C M, Expression and inheritance of multiple transgenes in rice plants, *Nature Biotechnol*, 1998 **16** 1060–4.

99. YODER J I and GOLDSBROUGH A P, Transformation systems for generating marker-free transgenic plants, *Bio/Technol*, 1994 **12** 263–7.

100. KOHLI A, GAHAKWA D, VAIN P, LAURIE DA and CHRISTOU P, Transgene expression in rice engineered through particle bombardment: molecular factors controlling stable expression and transgene silencing, *Planta*, 1999 **208** 88–97.

101. CHAREONPORNWATTANA S, THARA K V, WANG L, DATTA S K, PANBANGRED W and MUTHUKRISHNAN S, Inheritance, expression, and silencing of a chitinase transgene in rice, *Theor Appl Genet*, 1998 **98** 371–8.

102. KUMPATLA S P and HALL T C, Recurrent onset of epigenetic silencing in rice harboring a multi-copy transgene, *Plant J*, 1998 **14** 129–35.

103. KOHLI A, LEECH M, VAIN P, LAURIE D A and CHRISTOU P, Transgene organisation in rice engineered through direct DNA transfer supports a two-phase integration mechanism mediated by the establishment of integration hot spots, *Proc Natl Acad Sci USA*, 1998 **95** 7203–8.

104. McCABE M S, POWER J B, de LAAT A M M and DAVEY M R, Detection of single-copy genes in DNA from transgenic plants by non-radioactive Southern

blot analysis, *Mol Biotechnol*, 1997 **7** 79–84.

105. DUNWELL J M, Transgenic crops: The next generation, or an example of 2020 vision, *Ann Bot*, 1999 **84** 269–77.

106. DUNWELL J M, Transgenic approaches to crop improvement, *J Exp Bot*, 2000 **51** 487–96.

107. BOULTER D, Scientific and public perception of plant genetic manipulation – a critical review. *Crit Rev Plant Sci*, 1997 **16** 231–51.

108. ROGERS H J and PARKES H C, Transgenic plants and the environment, *J Exp Bot*, 1995 **46** 467–88.

109. McBRIDE K E, SVAB Z, SCHAAF D J, HOGAN P S, STALKER D M and MALIGA P, Amplification of a chimeric *Bacillus* gene in chloroplasts leads to an extraordinary level of insecticidal protein in tobacco, *Bio/Technol*, 1995 **13** 362–5.

110. HARDING K and HARRIS P S, Risk assessment of the release of genetically modified plants: A review, *Agro-Food-Indust Hi-Tech*, 1997 **Nov/Dec** 8–13.

111. VASIL I K, Molecular improvement of cereals, *Plant Mol Biol*, 1994 **25** 925–37.

112. SABRY S R S, SALEH S A, BATCHELOR C A, JONES J, JOTHAM J, WEBSTER G, KOTHARI S L, DAVEY M R and COCKING E C, Endophytic establishment of *Azorhizobium caulinodans* in wheat, *Proc Roy Soc B*, **264** 341–6.

4

Product development in cereal biotechnology

D. McElroy, Maxygen Inc., Redwood City

4.1   Introduction

During the brief history of modern cereal biotechnology, the development of innovative products has captured a great deal of academic interest and corporate attention, with such innovation being viewed as a highly visible source of competitive advantage. However, while there has been extensive basic research on the problems of characterising genes with specific agronomic effects, very little has been written about the problems associated with product commercialisation in cereal biotechnology. The increasing complexity of transgenic plant products, especially those containing metabolically engineered output traits,[1] will raise the cost of launching the next generation of products in cereal biotechnology. Then, more than ever, firms involved in cereal biotechnology will have to pursue two seemingly incompatible goals: increasing product differentiation while reducing manufacturing costs and development times in an effort to commercialise effectively their products.

4.1.1   Product development versus process development in cereal biotechnology

Product development focuses primarily on designing and testing prototypes of the product. Process development can be described as a system for creating and refining an organisation's capability to manufacture a product or set of products commercially.[2] While there is a large body of literature on the competitive advantage of efficient process development in mature industries, such as bulk chemicals, and an increasing focus on 'design for manufacturability' in assembly products such as automobile production, little has been published on the competitive role of effective process development in agricultural biotechnology.

Within cereal biotechnology, decreasing process development lead times should result in accelerated product development lead times, especially where innovations in process design are implemented on the road towards transgenic product launch. Rapid and effective ramp-up is essential for lowering the need for capital investments in breeding capacity, lowering current production costs, generating sales revenue and recovering development costs. Some of the resulting reduction in manufacturing costs can be passed on to farmers and other consumers to increase end-user acceptance of new products. Finally, innovative process design technologies can extend the proprietary position of a specific cereal biotechnology if would-be imitators or competitors are unable to determine how to produce the transgenic product either at competitive costs or at sufficient quality levels.

In this chapter I would like to describe how innovative process development, leading to rapid product commercialisation, can generate a competitive edge in cereal biotechnology.[3] A review of the overall product R & D cycle for developing transgenic cereals, using transgenic corn commercialisation in the USA as an example, should help illuminate how and where various process development activities fit into the broader product development efforts within cereal biotechnology. The product R & D cycle for transgenic plants can be divided into a number of stages that include: discovery, small-scale greenhouse efficacy screening and evaluation, the generation of intellectual property protection and regulatory approval, large-scale field evaluation and breeding and product release. The timing of these various stages for a hypothetical transgenic corn product is outlined in the project management chart shown in Fig. 4.1.

Finally, towards the end of the chapter, I will describe the development of transgenic corn products expressing B.t. transgenes. This example should illustrate how the contribution of each step in the process development cycle leads to the final commercialisation of this particular product.

4.2   Commercial targets for cereal biotechnology

One of the major goals of discovery research in cereal biotechnology is to identify genes that safely and effectively generate commercial opportunities in agriculture. The challenge for product designers in cereal biotechnology firms is to close the gap between what product can be offered to the customer, i.e. technology push, and what the customer really wants, i.e. market pull.[4] Technology push is a result of novel product opportunities arising from, for example, advances in genetic engineering, genomics, information management systems and grain characterisation technologies, while market pull comes from increasingly sophisticated end-user and consumer demand for improved processes and products.[5] The product traits themselves can be broadly divided between a first generation of relatively simple input traits, some of which are

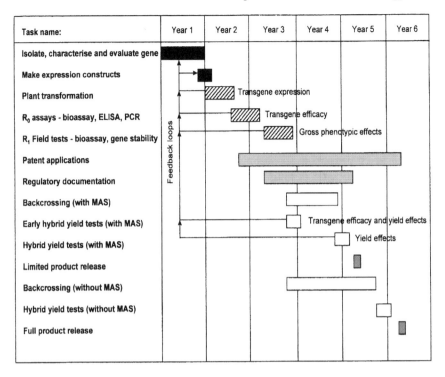

Task name:	Year 1	Year 2	Year 3	Year 4	Year 5	Year 6
Isolate, characterise and evaluate gene						
Make expression constructs						
Plant transformation		Transgene expression				
R_0 assays - bioassay, ELISA, PCR		Transgene efficacy				
R_1 Field tests - bioassay, gene stability		Gross phenotypic effects				
Patent applications						
Regulatory documentation						
Backcrossing (with MAS)						
Early hybrid yield tests (with MAS)				Transgene efficacy and yield effects		
Hybrid yield tests (with MAS)				Yield effects		
Limited product release						
Backcrossing (without MAS)						
Hybrid yield tests (without MAS)						
Full product release						

Fig. 4.1 Project management chart outlining the timing of the various development stages for a hypothetical transgenic corn product. The different steps in the product development life cycle are grouped as discovery (dark boxes), early generation, small-scale trait evaluation (striped boxes), intellectual property and regulatory filings (stippled boxes), late generation, large-scale field evaluation and trait breeding (light boxes) and product release (grey boxes). Important feedback loops that should exist between the various steps are shown. Approximate time lines for the development of a hypothetical corn product are indicated. MAS = Marker-Aided Selection. Adapted from McElroy.[3]

already on the market, and a second generation of increasingly complex output traits, some of which are about to come onto the market or are in the product pipeline.

Input traits include genes for pest and disease tolerance,[6,7] such as B.t. genes, α-amylase inhibitors, viral coat proteins, and viral replicase genes; herbicide resistance,[8] including those derived from mutation screening, e.g. resistance to sethoxydim (Poast®) and transgenic approaches, e.g. resistance to glyphosate (Roundup®) and phophinothricin (Liberty®); yield stability, including drought tolerance genes,[9] such as barley *HVA1*; and male sterility genes[10] for the generation of hybrid seed and for intellectual property protection. Most of these input trait products generate revenue by lowering the costs, financial and/or environmental, of plant production.

Output traits include genes for commercially valuable oils, proteins and starches; fatty acid modification; modification of seed storage proteins and amino acid profiles; and the manipulation of carbon-partitioning for novel starch production.[11] Efforts to minimise financial risk in agriculture have encouraged vertical integration, contract growing and end-product orientation with an associated focus on the measurement and categorisation of seed quality characteristics.[5] Most of these output trait products generate revenue by altering the specific measurable properties of the final product to meet these increasingly sophisticated consumer demands.

4.3   Problems in cereal biotechnology

There are high investment costs associated with the long lead time (6–10 years) for cereal biotechnology products to reach the marketplace and start generating revenue, and this is especially true for the output traits, such as those involved in the modification of grain quality. For such output products the impact on final agronomic performance of the transgene, and/or the effects of the plant genomic DNA flanking the transgene, might not be fully evaluated until well (3–6 yrs) into the breeding process (Fig. 4.1). There are a number of other risks associated with cereal biotechnology products that need to be borne in mind during the initial design phase. There can be uncertain profitability associated with some of the 'technology push' concepts at the onset of product development. There may be intellectual property issues limiting a company's freedom to operate with key technologies or encouraging the rapid development of potentially superior technologies by their competitors. Finally, there might be uncertainty associated with regulatory and consumer acceptance issues inhibiting trade and domestic investment, for example, the export of processed grain derived from transgenic plants to the EU.[12]

New products coming out of discovery research do not necessarily ensure lasting profits. Returns from innovation can be competed away unless isolating mechanisms are in place to inhibit imitators. As the mean development time to launch the next generation of output products increases, the time lag between gaining regulatory approval and patent expiration will inevitably shrink. Although patent protection lasts a number of years following the date of initial filing, some portion of this period will be lost because of the time required to 'fine tune' the technology, conduct efficacy trials, secure regulatory approval and produce the finished transgenic product in sufficient sales volumes. Instead of conducting their own lengthy and costly regulatory procedures, the producers of generic (off-patent) cereal biotechnology products, which could start coming onto the market in the next decade, might need only to show that their version of the product is chemically and biologically equivalent to the original patented version. Therefore, for off-patent products, firm-specific differences in their process knowledge base will become an increasingly valuable source of competitive advantage.

4.4 Efficacy screening of commercial traits

The development of a cereal biotechnology process plan begins with the design of those analytical methods, screens and other assays (Table 4.1) that will be used to detect the expression of the desired gene in early generation plants.[4,13] However, the evolution of analytical methods should continue throughout the project, with later development emphasising methods applicable to production quality control during breeding and foundation seed production (Fig. 4.2, Table 4.2). Analytical techniques play a critical role in evaluating process R & D experiments throughout the product development life cycle by helping to generate a deeper understanding about underlying cause-and-effect relationships

Table 4.1 Molecular technologies commonly used in transgenic plant analysis

Technology	Application	Generation*
Transgene-specific PCR	High throughput transformant screen, high throughput transgene segregation, regulatory approval, e.g. Amp^-.[†]	R_0/R_1 R_1/BC_1 BC_1
Event-specific PCR	High throughput QC in backcrossing.	BC_n
Quantitative PCR	Copy number determination, zygosity determination.	R_0/R_1 S_n
Southern analysis	Transgene complexity/copy #/fingerprint, multi-locus transgene segregation, regulatory approval, e.g. Amp^-.	R_1, BC_n R_1 BC_n
RT-PCR	Semi-quantitative transgene expression, regulatory approval, e.g. Amp^-.	R_1 BC_n
Northern analysis	Transgene expression, transcript characterisation, e.g. size, regulatory approval, e.g. Amp^-.	R_1 R_1 BC_n
Western analysis	Transgene expression, transgene product processing, quantity, regulatory approval, e.g. Amp^-.	R_0/R_1 R_0/R_1 BC_n
Quantitative ELISA	High throughput transgene expression, high throughput transgene product quant.	R_1 R_1/BC_n
Qualitative ELISA (DipStick)	High throughput transgene expression.	BC_n/S_n
Quantitative analytical	Transgene effect, e.g. lysine by HPLC.	R_1–S_n
Phenotypic assay	High throughput transgene expression, e.g. herbicide resistance.	R_0–S_n

* See Fig. 4.2.
† Amp^- = free of any bacterial selectable marker gene, such as that encoding resistance to ampicillin.

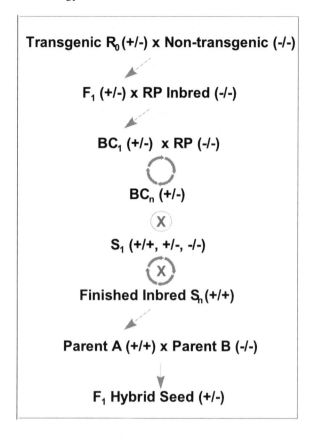

Fig. 4.2   Outline of the breeding process for a transgenic corn product. The genotypes of the various generations are indicated as either present (+) or absent (−) for the transgenic trait. Abbreviations used include: RP, recurrent parent inbred line; BC, generation resulting from a backcross; X, selfcross; S, generation resulting from a selfcross. A rolling circle with arrows indicates an iterative process. Adapted from McElroy.[3]

in the recombinant trait's efficacy. Particularly novel products will require the invention of novel analytical techniques, tools and procedures. The ability to recognise the need to develop novel analytical techniques in a timely manner is an important capability within any product commercialisation group.

Screening methods generally fall into two major types, selective screens or analytical screens. The development of selective screening systems, e.g. for herbicide tolerance, may begin in tissue culture, followed by initial greenhouse trials and experimental fields where promising transgenic lines are evaluated in randomised, small-scale plots. The development of such selective screening systems has been important in the evaluation of input trait efficacy, where many of the products can be evaluated in early generations (Fig. 4.1) in genotypes that are readily transformed but are generally not commercial varieties. Analytical

Table 4.2   Corn backcross conversion – effects of marker-aided selection

Period	Activity	Product*	Trad.	Marker-aided
			Mean % RP[†]	Mean % RP[‡]
Yr1	Backcross (BC) R_0 transgenics to			
	elite inbreds (recurrent parents, RP)	F1	50.00	50.00
	Select transgene – BC to RP	BC1	75.00	82.99
	Select transgene – BC to RP	BC2	87.50	97.89
Yr2	Select transgene – BC to RP	BC3	93.75	~100.00
	Select transgene – BC to RP	BC4	96.87	–
	Select transgene - BC to RP	BC5	98.44	–
Yr3	Select transgene – BC to RP	BC6	99.22	–
	Select transgene – BC to RP	BC7	99.61	–
	Self	S1	99.61	~100.00
Yr4	Self	S2	99.61	~100.00
	Identify transgene expressing line			
	phenotypically similar to RP	S3	99.61	~100.00
	Testcross	S3 TC	49.81	~50.00

* See Fig. 4.2.
† $E[p(n)] = [1 - (1/2)^{n+1}] \times 100\%$, with associated variance and S.D. a function of genetic map length and no. of chromosomes.
‡ Identify plants carrying the transgene prior to tissue sampling for efficient selection.

screening systems, e.g. for transgene presence, event verification, oil content, amino acid composition and other grain quality characteristics, may be followed by a subsequent functional assay if such an assessment is critical for the initial identification of efficacious lines. For those output traits affecting grain quality characteristics this may involve small-scale compositional analysis followed by large-scale animal feeding trials or large-scale processing experiments. Many grain quality characteristics can be fully evaluated only in the finished inbred line (Fig. 4.1) either to be sold directly or used in hybrid seed production. The generation of such finished lines is both time consuming and expensive.

During the early generation evaluation phase of the product development cycle, the company's goal is to collect sufficient data on the performance of the candidate gene to warrant the much more expensive step of large-scale field evaluation and entry into the breeding program. At this early stage gene candidates can be abandoned and/or targeted for redesign as a result of either a lack of demonstrated efficacy, gross agronomic abnormalities or phytotoxicity (Fig. 4.1).

The precise principles governing the expression of recombinant traits in transgenic plants are poorly understood. As a result, problem solving in cereal biotechnology relies more heavily on physical experimentation than on conceptual modelling. Thus, while early greenhouse observations are critical, the real uncertainty lies in predicting a recombinant trait's agronomic performance in elite commercial germplasm. This is especially true for output traits the efficacy of which can be significantly influenced by genotype effects. In this environment one tends to find that 'learning by doing' is the only available process development strategy.

4.5   Molecular breeding of transgenic plants

The aim of the breeding program is to develop high-yielding competitive varieties free of any yield drag or other agronomic problems that might be associated with either the novel gene's expression or its initial genetic background.[14] This breeding process might take 5–9 years to complete in corn, depending upon the number of backcrosses involved (Fig. 4.2, Table 4.2). Once early (R_0/R_1) generation transgenic lines have been characterised they will be entered into a breeding program with a number of quality inbred lines to do initial assessments of genotype and environmental effects on the performance and genetic stability of the new trait (1–2 years). The efficacy of the novel trait must be assessed through several seasons in multiple locations to ensure that performance standards are consistently met in a number of unique environments. Those transgenes and transgenic events that make it through this initial field evaluation will eventually enter into a variety development program, involving crossing into other high-quality lines (3–5 years). Towards the end of varietal development, seed production begins to generate enough seed stocks for commercial release through normal foundation seed multiplication channels (1–2 years). Individual transgenic 'events' are eliminated at each step in the breeding program as they fail to meet performance benchmarks, so the initial evaluation process must start with a large number of high-quality transgenic lines.

4.5.1   Shortening breeding programs

A number of processes can be used to shorten transgenic breeding programs. The development of transformation protocols for cultivars containing a genetic background common to that of elite commercial varieties will reduce the number of backcross generations needed to produce a fully converted transgenic inbred line. Winter nursery facilities can be used for year-long production of multiple product generations. Double-haploid plant production can be used to produce homozygous transgenic lines for rapid evaluation of transgene copy number effects, yield drag associated with individual insertion events or combining ability for stacked products containing more than one transgene. Finally, marker-aided selection can be employed to facilitate greatly the selection of progeny for use in subsequent backcrossing to recurrent parent lines.

4.5.2   Marker-assisted selection for transgenic plants

Valuable alleles and/or transgenes can be tracked in breeding populations using genetically linked molecular markers.[14,15,16] A number of marker systems are available for this process including: restriction fragment length polymorphisms (RFLPs); simple sequence repeats (SSRs); amplified fragment length poly-morphisms (AFLPs); and single nucleotide polymorphisms (SNPs). Marker-aided selection (MAS) is used to select for genetic similarity to the targeted elite inbred (recurrent parent) within early backcross generations. MAS facilitates the elimination of a number of generations from the backcross conversion process and might save 1–2 years in transgenic product development (Table 4.2). MAS also allows one to select for less linkage drag associated with the 'non-elite' genomic DNA flanking the transgene (Fig. 4.3), increasing the probability of obtaining an acceptable conversion. It is advisable to use a sufficient number of markers to sample the genome adequately, with the actual number depending upon the genetic relationship of the transgenic line to the recurrent parent, the backcrossing strategy and the breeder's experience. Uniformly distributed markers are more effective in backcross conversion and minimising linkage drag than an equal number of markers per chromosome or randomly assigned markers.[10] Using MAS to accelerate backcross conversion is critical to the introduction of value-added traits in elite genetic backgrounds, the efficient management of seed inventories and the timely commercial launch of new transgenic products.

4.6   Molecular quality control for transgenic plants

Transgenic plant lines containing independent transgenes and independent transgenic insertion events are backcrossed and evaluated at a number of different breeding locations. This is done, in part, to evaluate transgene efficacy under a number of different environmental conditions. Within this breeding and evaluation process there exists the potential for line and/or gene misidentification.[17] It is important that events should not be mixed up in the breeding process since only one event might be deregulated for eventual sale. Finally, at the end of the breeding process, the customer must be provided with an estimate of the quality of the transgenic product that they are purchasing, where quality is represented by the percentage of 'off types' they can expect in their field of transgenic plants.

In breeding programs transgenic lines are often identified by screening for the presence of a more readily detectable selectable marker gene (e.g. herbicide resistance) that was co-introduced with the transgene of interest. Insertion event-specific transgene silencing, recombination-mediated transgene deletion and/or pollen 'contamination' may result in loss of expression for the non-selected transgene of interest. Without a quality control system in place, generations of selection for the readily detectable marker gene may pass before the presence/efficacy of the transgene of interest is evaluated. Therefore, independent

(a) Molecular breeding: BC$_1$ line production

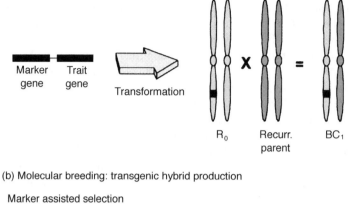

(b) Molecular breeding: transgenic hybrid production

Marker assisted selection

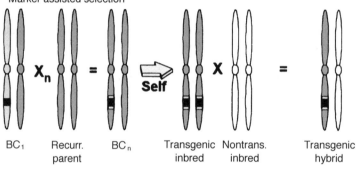

Genetic marker spread = 1 every 15cM, 1cM = 1.2 Mbp in corn
non-elite gDNA ~ 7.5 cM ~ 9 Mbp

Fig. 4.3 Overview of breeding a transgenic corn product. (a) Plant transformation and the production of the first backcross generation (BC$_1$) following crossing of the R$_0$ plants with a recurrent parent inbred line. The transgene event ■ is shown integrated into the hemizygous R$_0$ plant chromosome ▨ with the homologous chromosomes for the recurrent parent shown in ▨. (b) Production of the transgenic hybrids. Following several rounds of backcrossing and selection with the recurrent parent a final backcross generation (BC$_n$) is produced. This hemizygous BC$_n$ generation is selfed, homozygous transgenic lines are identified and the homozygous material bulked up in foundation seed nurseries. The homozygous transgenic lines are then crossed with other elite inbred lines to generate the transgenic hybrid seed products that will be sold. Adapted from McElroy.[3]

transgenes and transgenic insertion events must be tracked and verified throughout the breeding and evaluation process to ensure that transgenes are not 'lost' or transgenic lines misidentified. This gene tracking process[13] has been dependent upon the development of a number of analytical tools (Table 4.1), including rapid transgene and insertion event-specific assays. The gene tracking process has also been facilitated by developments in the automation of DNA isolation/analysis and in data processing.

The recent evolution of quality control systems in cereal biotechnology is a good example of prior production experience, or 'learning by doing', generating data that can contribute to future process development projects, thus allowing for 'learning' to take place 'before doing' subsequent process development projects. In relatively immature industries, such as cereal biotechnology, manufacturing performance should improve with cumulative experience. Cumulative production experience generates data for systematic problem identification and problem solving, leading to improvements in product design, equipment modifications and worker training. However, for such 'learning by doing' to be retained and utilised, there need to be strong feedback loops from the production process (in this case the breeders) to process R & D teams (Fig. 4.1).

4.7   Intellectual property and freedom to operate

Freedom to operate (FTO) under any dominating patents must be secured in order to commercialise a product.[18,19] Transgenic plants often utilise several patented technologies, each owned by different parties, e.g. transformation method, selectable marker gene, specific genes, and gene expression elements. Therefore, companies must commit substantial resources to develop either in-house, or acquire rights to utilise, all of the technologies used in the development of their commercial products. These issues should ideally be addressed before the product enters the breeding program for large-scale field evaluation.

In order to obtain a US patent, the technology must be novel (no prior art), non-obvious (no analogous prior art) to one ordinarily 'skilled in the art', and have some defined utility. The conception and reduction to practice of the claimed invention (the actual invention and any modifications to the invention that retain the same activity that are taught in the patent application) should be described in such a way that anyone in the field could make and use the invention. From a product development point of view in cereal biotechnology patents may be of essentially two major types. A product patent covers compositions of matter such as genes and gene products, monoclonal antibodies, engineered cells, and transgenic plants. A process patent covers methods such as those for gene cloning, transformation, gene silencing, gene delivery, and molecular breeding.

There are a number of problems associated with patents with respect to process development. First, there is usually a long delay between an initial patent application and the issuing of the patent. In the US, the contents of a patent application are not released until the patent is actually issued. Second, there are often a number of overlapping claims between related patents and there is often inconsistency in the breadth of coverage granted between different countries. Third, there can be significant resource requirements to file the patents and, especially in cereal biotechnology, to enforce the patent. Finally, overly broad patent protection is normally not granted at the present time. Such broad patents are generally not written in such a way as actually to teach anyone skilled in the art how to repeat all of the application's claims and the granting of such broad patent protection is also believed to inhibit subsequent innovation.

All of these patent issues, while problematic from the biotechnology firm's point of view, have been a boon to patent litigation lawyers. Thankfully there are a number of solutions to these patent problems that should be investigated during the product's development. Licensing and cross-licensing of technology between firms can occur in order to exchange rights to patents critical to product commercialisation. Mergers and acquisitions, either partial or complete, can be orchestrated in order to gain patent rights for critical technological component(s) in the commercial product. Finally, firms can engage in technology development to supersede a competitor's patents.

4.8   Regulatory issues and risk assessment

There is a widespread belief that standard plant-breeding procedures are generally not adequate to identify all the potential problems associated with transgenic plants.[12] It is argued that, as a consequence of the diversity of available transgenes, the environmental consequences of genetic modifications cannot reasonably be anticipated. Therefore, countries have established their own agencies to oversee the introduction of new transgenic plant varieties into their environments and the importation of novel plant products, particularly transgenics, into their marketplaces.[20,21] In the establishment of these national agencies, most governments have striven to balance the need to develop oversight mechanisms that address potentially serious risks without generating a regulatory burden which creates disincentives to innovation and diminishes the availability of important new techniques and products.

Securing government regulatory approval for product introduction and addressing customer acceptance issues are critical aspects of product development and commercialisation, e.g. for key export markets. In addition to gaining approval of the transgenic 'event' itself, e.g. for herbicide resistant crops, the registration and deregulation of any associated chemicals used with the transgenic varieties may also be required. In most countries there are two kinds of regulation that govern research and development of transgenic plants. Rules for 'contained use' govern genetic modification in the laboratory, concentrating mainly on worker health and safety issues. In the US the rules for 'field release' focus on environmental risk assessment appropriate to the nature and final use of the transgenic plant, with each release initially considered on a case-by-case basis in order to build up experience with particular crop/transgene combinations.

A number of factors are borne in mind during the risk assessment process associated with transgenic plant deregulation.[20,21] These factors include: the function of the gene in the donor organism; the effect of the transgene on the phenotype of the transgenic plant; the risk of the transgenic to animals and humans, e.g. evidence of toxicity and/or allergenicity; the ability of the transgenic to colonise and persist in agricultural habitats (weediness); and the proximity of the transgenic species to its 'centre of origin'. These last two factors address the likelihood and consequence of transgene movement to other weedy or wild relative plants by cross pollination or to other (pathogenic) organisms by horizontal gene transfer. Other risk issues include the impact of the transgenic plant on the evolution of target organisms, e.g. insect resistance to B.t.; the behaviour of non-target organisms, including minor pests or beneficial insects; the existing ecological relationships within the agricultural system, such as the plant's potential to disrupt existing pollination systems; and the ability of the transgenic to provide a resource for organisms that are pests of other crops within the agricultural ecosystem.

Individual nations generally employ one of two different kinds of regulatory systems, with the data assessed in both being similar. Horizontal regulatory systems, such as those within the EU, are process-based systems that apply to all plants produced by a certain method, e.g. by a particular transformation method. Vertical regulatory systems, such as that in the US, are product-based systems that define the characteristics of modified plants that require them to be regulated. So, except for FDA approval, the US has a techniques-based regulatory system which differentiates between 'natural' exotics, genetic improvements by conventional methods, and manipulation by recombinant DNA techniques.

Within the US, transgenic plants are regulated by three government agencies, following a 1986 decision to regulate transgenics under existing statutes. The US Department of Agriculture (USDA/APHIS) controls permits for inter-state movement and field release of transgenic materials, assesses the pest potential of transgenic plants and determines when transgenics can be field-grown without notification or permits, otherwise known as deregulation. The Food and Drug Administration (FDA) determines that a transgenic plant has been adequately evaluated in accordance with its biotechnology food and feed policy, e.g. for the safety of antibiotic selectable marker genes. Finally, the Environmental Protection Agency (EPA) regulates transgenic plants with pesticidal properties, e.g. B.t. plants, and determines pesticide/herbicide residue tolerance on transgenic plants. In Canada, transgenic plants are regulated by Ag. & Agri-Foods Canada and Health Canada. In the Canadian system, products are similarly regulated whether generated by mutation or transformation. This is essentially a risk-based system that focuses on the intrinsic properties of the product, e.g. weediness, invasiveness, or its potential for outcrossing, rather than on the method of production of the variant under evaluation. In Europe, a Novel Foods Regulation went into effect in 1997 to replace completely EU Directive 90/220 that previously covered environmental issues associated with transgenic seed sales and transgenic grain import. In Japan, transgenic plants are regulated by the Ministry of Agriculture, Forestry and Fisheries and the Ministry of Health and Welfare. To date, there is no international harmonisation of regulations to ensure that transgenic plant varieties released in one country will be accepted in another, so antibiotic resistance genes in food products might inhibit international trade in transgenic products. Meanwhile, The Convention on Biodiversity has established a biosafety working group to develop safety standards for international trade in transgenic products.[22]

4.9   Product release and marketing strategies

In order to recover its substantial R & D investments, the developer of a commercial product might either generate and market the transgenic seed directly or negotiate a royalty with a seed company/companies, e.g. for technologies discovered by universities, government agencies, technology development companies or large agrochemical companies. Because of the previously described learning curve effects in process development, new products might have higher manufacturing costs than older products. Thus the profit margins on a newly launched product might be small or even negative during the first few years of its commercial life.

In transgenic plants, the product is carried in a vegetatively or sexually reproduced form which the customer could potentially propagate following its initial purchase. For hybrid crops, new seed must be purchased each year; therefore the transgene is protected and premiums for the technology can be added to the seed. However, for inbred or vegetatively propagated crops, where transgenic material could be replanted each year, the transgene is not biologically protected; therefore farmers might be obliged to take a licence from the seed company that allows them to buy and grow seed containing the transgene. Alternatively, firms might need to develop biological processes to protect their investment, such as the so called 'terminator' technology.[23]

The marketing strategy for any one particular trait will be influenced by the nature of the product. Input traits, such as those conferring herbicide, disease or stress tolerance, which enhance yield or reduce inputs, can return revenue either by increasing the seed company's market share and/or increasing seed sale premiums. For some products, such as herbicide tolerant crops, the agrochemical companies that helped develop these products can benefit from increased chemical sales, e.g. Roundup-Ready® Corn or Liberty-Link® Corn. For output products, such as improved grain quality, there may be either a seed sale premium, an end-product premium for contract-grown crops or a mechanism for sharing the added value with the end-user, such as a food-processing company.

4.10   Product development: a practical example

A review of the overall product R & D cycle for developing transgenic B.t. corn, should help illuminate how and where the previously described product and process development activities fit into the broader product commercialisation efforts within cereal biotechnology.

Table 4.3   Spectrum of activity of B.t. genes

B.t. gene	Spectrum of activity	Transgenic product
CrylAb	European corn borer, southwestern corn borer, corn earworm, fall armyworm	Corn
CrylAc	Tobacco budworm, cotton bollworm, European corn borer, southwestern corn borer, corn earworm, fall armyworm	Cotton Corn
Cry9c	European corn borer, southwestern corn borer	Corn
Cry3A	Colorado potato beetles	Potato

4.10.1   B.t. genes, proteins and mode of action

Bacillus thuringiensis bacteria produce insecticidal proteins, each with their own spectrum of activity (Table 4.3). These accumulate as crystals and are thought to aid in bacterial spore germination and vegetative growth following ingestion and gut wall destruction in certain vector organisms. B.t. protein specificity is determined by its interaction with aminopeptidase-N receptor molecules on the luminal side of the insect midgut epithelium. B.t. protein toxicity appears to be determined by the formation of a pore in the cell membrane, which leads to cell lysis and disintegration of the midgut epithelium. The B.t. crystal protein consists of three domains: a helices responsible for pore formation; a receptor binding domain; and a third domain of unknown function. By exchanging corresponding domains between crystal proteins their insecticidal activity can be exchanged, expanded and/or improved.

There are a number of potential advantages associated with the development of transgenic B.t. crops. Their development and deployment will reduce the cost, time and effort spent protecting crops from insects and, relative to many commercially available chemical insecticides, they should contribute to an environmentally friendly crop production system.

4.10.2   Development of transgenic crops expressing B.t. transgenes

B.t. corn with improved European corn borer (ECB) resistance was first released to farmers in 1996 and a number of competing B.t. corn products are now on the market (Table 4.4). However, a number of product design and process development issues had to be addressed in order to develop and commercialise transgenic crops expressing B.t. transgenes.[24]

Efficient transformation and selection systems had to be developed for all major crops with the choice of transformation protocol used by any one company being influenced by intellectual property considerations. Efficient transgene expression systems needed to be generated. These included the use of synthetic B.t. genes optimised for plant codon preference, truncated B.t. genes,

Table 4.4 B.t. corn events registered with the EPA and 1998 seed market share estimates

Company	B.t. gene	Transgene expression	B.t. event	Trademark	Est. 1998 acres, %
Monsanto	*CryIAb*	Constitutive	MON810	YieldGard®	76
Novartis (NK)	*CryIAb*		BT11	YieldGard®	
Novartis (Ciba)	*CryIAb*	Green tissue	176	Maximizer®	17
Mycogen	*CryIAb*	& pollen	176	NatureGard®	
DEKALB	*CyIAc*	Constitutive	DBT418	Bt-Xtra®	5
AgrEvo	*Cry9c*	Constitutive	PGS351	Starlink®	1

and highly active constitutive promoters for driving high level expression in transgenic plants.

Companies had to establish efficient transgenic plant production systems. Gene transfer can result in multiple transgene integrations and rearrangements of the heterologous and/or homologous DNA with associated effects on transgene expression. In many cases a large number of transgenic events needed to be produced in order to identify the few lines with adequate transgene expression levels and agronomic performance. Transgenic plant identification and characterisation methods had to be established and validated. Molecular breeding and molecular quality systems had to be put in place to handle transgenic products.

New regulatory systems for transgenic plants had to be established. For example, in the US the EPA considers B.t. crops as pesticides and they need to be registered as such, including toxicological studies on a range of organisms. Companies hoping to sell B.t. products needed to design and implement resistance management (RM) strategies, with grower recommendations and monitoring protocols to ensure that resistant insects did not evolve so that the lifetime of the product could be maintained. Systems had to be developed to integrate these new technologies into current farming practices. Finally, there was, and continues to be, a need to establish acceptance of transgenic crops and crop products by growers, food processors and consumers.

4.10.3 Insect resistance management

The success of B.t. crops will depend on whether target pests develop resistance to them. Specific insect resistance management (IRM) measures have been, and are continually being, designed and implemented to minimise the build-up of resistance genes in insect populations that might develop under constant selection pressure from transgenic crops.[25] A number of different IRM strategies have been proposed. In the 'gene stacking strategy' transgenic plants are produced that express a combination of different B.t. genes, or other insect resistance genes, each with a unique mode of action that should delay the evolution of resistant insect populations. In the 'high dose + refuge strategy' transgenic plants are developed with a *high dose*, or relatively high level of B.t. protein accumulation, to eliminate resistance genes in heterozygous resistant insects that are only slightly less susceptible to B.t. than fully susceptible insects. At the same time a *refuge* of non-B.t. plants is provided to dilute resistance development in insect populations. In 1998 Monsanto's IRM Strategy for their YieldGard® B.t. corn allowed growers either to plant at least 5% of their acres with a non-B.t. corn hybrid and not treat the rest with insecticides registered to control ECB or to plant at least 20% of their acres with a non-B.t. corn hybrid and treat all their corn acres with non-B.t. insecticides as needed.

4.10.4   Commercial goals of insect resistant corn product

The goal of insect resistant corn is to redirect a substantial portion of the insecticide market from agrochemicals to the seed industry. Almost 25% of all pesticides used in US agriculture are insecticides. In 1996 the insecticide market was estimated to be ~$7 billion. However, for transgenic plants expressing B.t. genes, it should be borne in mind that their initial ECB target is a cryptic feeder that spends most of its time in the stalk and so it is a pest that is inaccessible to most conventional insecticides. Therefore, it is uncertain exactly how much of the conventional insecticide market has been captured by B.t. plant products.

The B.t. technology itself is an insurance product, with the return to the farmer being directly proportional to the level of potential yield loss that would have been caused by insect damage in the absence of any protection afforded by the B.t. transgene. The break-even point is estimated to fall between 2–4% potential yield loss, depending upon the yield potential of the crop (Table 4.5). During 1997 the actual yield advantage for B.t. corn over non-B.t. corn in the mid-west US was determined to be between $15 and $43 bushels/acre (Table 4.6).

From the onset B.t. corn was expected to increase yield by ~5% in ~20% of the 82 million acres of hybrid corn planted in 1998, representing a potential gross added value of ~$350 million to those who developed this transgenic cereal product. B.t. corn sales in 1998 represent an added gross premium of ~$150 million (Table 4.7), with the final net added value being dependent upon

Table 4.5 Added value* per acre from using B.t. hybrid: % yield protection by preventing ECB damage

Yield potential (bu/a)	0%	2%	4%	8%	16%
150	−$10.94	−$3.44	$4.06	$19.06	$49.06
200	−$10.94	−$0.94	$9.06	$29.06	$69.06

* Assume B.t. seed premium is $35/unit (80,000 kernels), one unit plants 3.2 acres (25,000 kernels per acre), corn price at $2.50/bu.

Table 4.6   1997 yield advantage* for B.t. vs. non-B.t. corn isolines (source: Monsanto)

State	Yield advantage B.t. vs. non-B.t. (bu/A)	Yield advantage B.t. vs. non-B.t. ($/A)
Iowa	21.6	$43.06
Illinois	17.4	$32.56
Nebraska	14.7	$25.81
Indiana	11.9	$18.81
Ohio	6.3	$15.75

* Assume B.t. seed premium is $35/unit (80,000 kernels), one unit plants 3.2 acres (25,000 kernels per acre), corn price at $2.50/bu.

Table 4.7   US B.t. corn seed company market share estimates, 1998 (source: Michael Judd, Paine Webber)

Company	Acres, millions	Acres, %	Revenue $ millions
Pioneer	7.2–7.8	44	75
Novartis	6.0–7.0	38	39
DEKALB	1.2–1.4	7	13
Mycogen	0.3	2	2
Others (13)	1.3–1.5	9	15
Total	16–18	100	150

the costs associated with seed production, quality control, royalties to third-party technology providers, market share gain, discounting, dealer costs, etc.[26]

4.11   Future trends

Efforts to minimise financial risk will continue to encourage functional consolidation, vertical integration, contract growing and end-product orientation in modern commodity agriculture. The resulting agribusiness conglomerates should have a number of common enabling features in order for them to take full advantage of future product commercialisation opportunities in cereal biotechnology. They should have access to proprietary, elite germplasm so that they can piggy-back their traits onto already high-yielding varieties. They should have the ability rapidly to move traits into finished products with an extensive, efficient seed production, distribution and sales system. They should have a significant market presence (sales > 150,000 units/yr) in order to capitalise fully on the investment, direct or indirect, required to obtain the transgenic trait. Finally, they need to have innovative customers who are capable of responding to new product opportunities. From the description of the product R & D cycle in cereal biotechnology, one can also conclude that a major challenge facing firms today is to create development processes that are fast, efficient, and capable of generating high-quality products. While process development can have a significant impact on a product's manufacturing cost, savings can be missed by failure to invest adequate resources in process development early enough in the product life cycle.

It should now be apparent from the foregoing description that development projects have two major outputs, the technology that is implemented in the new process and organisational knowledge that becomes available for future projects. Organisation knowledge includes how to manage projects, allocate resources, assign personnel and resolve disputes. Each new project should generate feedback (Fig. 4.1) on gaps between expected and desired lead times, uncover critical parameters, and create insights about the process performance that triggers a search for new approaches to managing future development projects. Much of this new knowledge becomes embedded in the firm's organisational routines and procedures. However, high performance in process development requires the ability to learn, and some organisations will be better able to learn from their experiences than others. Learning is rooted in how organisations feed data from manufacturing (breeding, foundation seed and sales) back into its technical knowledge base allowing R & D to anticipate future problems. Organisational or geographical barriers between process R & D and plant breeding can have a significant negative consequence if R & D scientists are incapable of designing processes that work well in the field.

In conclusion, the success of any individual organisation will depend upon its ability to do more proficiently than their competitors those things that their products demand and their customers value. As cereal biotechnology matures, and consumer prices eventually stabilise, the ability to develop rapid and efficient processes will be an increasingly important source of competitive advantage in the commercialisation of products in cereal biotechnology.

4.12   References

1. HERBERS K, SONNEWALD U, 'Manipulating metabolic partitioning in transgenic plants', *TIBTECH*, 1996, **14** 198–205.
2. PISANO G P, *The Development Factory*, Boston, Harvard Business School Press, 1997.
3. McELROY D, 'You snooze, you lose: racing to market in ag biotech', *Nature Biotech*, 1999, **17** 1071–4.
4. MAZUR B J, 'Commercializing the products of plant biotechnology', *TIBTECH*, 1995, **13** 319–23.
5. COOK M L, 'Structural changes in the U.S. grain and oilseed sector in food and agricultural markets', In *The Quiet Revolution*, Schertz and Daft (eds), National Planning Association, 1994.
6. DIEHN S H, DE ROCHER E J, GREEN P J, 'Problems that can limit the expression of foreign genes in plants: lessons to be learned from *B.t.* toxin genes', in

Genetic Engineering, Vol. 18, Setlow J K (ed.), Plenum Press, New York, 1996, 83–99.

7. GEBHARDT C, 'Plant genes for pathogen resistance – variations on a theme', *TIPS*, 1997, **2** 243–4.

8. DUKE S O, *Herbicide-resistant crops: agricultural, environmental, economic, regulatory, and technical aspects*, Lewis Publishers, 1996.

9. INGRAM J, BARTELS D, 'The molecular basis of dehydration tolerance in plants', *Ann Rev Plant Physiol Plant Mol Biol*, 1996, **47** 377–403.

10. HORNER H T, PALMER R G, 'Mechanisms of genic male sterility', *Crop Sci*, 1995, **35** 1527–35.

11. OHLROGGE J B, 'Design of new plant products: engineering fatty acid metabolism', *Plant Physiol*, 1994, **104** 821–6.

12. BOULTER D, 'Scientific and public perception of plant genetic manipulation – a critical review', *Crit Rev Plant Sci*, 1997, **16** 231–51.

13. REGISTER J C, 'Approaches to evaluating the transgenic status of transformed plants', *TIBTECH*, 1997, **15** 141–6.

14. HILL W G, 'Variation in genetic composition in backcrossing programs', *J Heredity*, 1993, **84** 212–13.

15. HILLEL J, SCHAAP T, HABERFIELD A, JEFFREYS A J, PLOTZKY Y, CAHANER A, LAVI U, 'DNA fingerprints applied to gene introgression in breeding programs', *Genetics*, 1990, **124** 783–9.

16. HOSPITAL F, CHEVALET C, MULSANT P, 'Using markers in genetic introgression breeding programs', *Genetics*, 1992, **132** 1199–210.

17. APEL A, 'Monsanto, Limagrain recall genetically engineered RR canola seed', *Seed & Crop Digest*, 1997, **6** 23.

18. CHAHINE K G, 'Going beyond the native: Protecting DNA and protein patents', *Nature Biotech*, 1997, **15** 1–2.

19. CHAHINE K G, 'Patenting DNA: Just when you thought it was safe', *Nature Biotech*, 1997, **15** 586–7.

20. BARTON J, CRANDON J, KENNEDY D, MILLER H, 'A model protocol to assess the risks of agricultural introductions', *Nature Biotech*, 1997, **15** 845–8.

21. DALE P J, 'R & D regulation and field trialling of transgenic crops', *TIBTECH*, 1995, **13** 398–403.

22. HOYLE R, 'The biosafety protocol slouches towards Montreal', *Nature Biotech*, 1997, **15** 694.

23. OLIVER M J, MELVIN J, QUISENBERRY J E, TROLINDER N L G, KEIM D L, 'Control of plant gene expression', US Patent, 1998, **5** 723–65.

24. PEFEROEN M, 'Progress and prospects for field use of Bt genes in crops', *TIBTECH*, 1997, **15** 173–7.

25. HUTCHINSON W D, OSTLIE K R, 'Bt corn and European corn borer: Long-term success through resistance management', North Central Region Extension Publication NCR 602, Iowa State University, Ames, Iowa, 1997.

26. CASH A W, JUDD M T, *BT-Corn mules, quarter horses and thoroughbreds* PaineWebber Co., New York, 1998.

5

Using biotechnology to add value to cereals

R. J. Henry, Southern Cross University, Lismore

5.1 Introduction

The value of cereal crops can be improved in two main ways. Firstly, the quantity of crop produced can be increased and secondly, the quality of the harvested grain can be enhanced. Linked to the question of quality is also the improvement of safety (Henry and Kettlewell 1996). These two approaches are not entirely independent. Improvements in the productivity of cereal varieties can be associated with improved quality of the grain.

Removing constraints imposed by biotic stress (the impact of pests and diseases) is an attractive option for improving the productivity of cereal crops. The introduction of genes conferring resistance to major pests and diseases provides an opportunity to improve productivity by removing the losses associated with the specific disease targeted. The introduction of herbicide resistance into cereal crops allows a reduction of losses associated with competition from weeds. Both these approaches may lead to associated improvements in grain quality. Freedom from weeds can reduce or eliminate the problem of weed seed contamination in cereal grain. Removal of disease constraints may result in improved grain quality avoiding the losses in quality associated with the presence of disease organisms in the crop. Diseases often result in reduced grain size and associated quality deterioration.

The quality of cereal grain may be improved either in a nutritional sense or by improving the processing properties of the grain. Most conventional plant breeding has addressed the need for appropriate processing qualities in new cereal varieties. Biotechnology may allow more emphasis to be placed on novel alterations of nutritional quality.

Molecular markers can be used to improve the efficiency of cereal breeding programs aiming to improve the value of cereal crops. Molecular analysis may also be applied in fingerprinting or identification of cereal genotypes with more immediate potential for improvement of cereals. For example, molecular analysis of genotypes can be used to monitor seed purity and identity prior to planting and to characterise grain lots in trading and processing (Henry *et al.* 1997). The composition of cereal-based foods or products can be monitored to ensure authenticity of labelling. These applications of biotechnology can have an almost immediate impact on the quality and value of cereal production.

Genetic engineering offers the possibility of going beyond these short-term outcomes of biotechnology applications to the generation of more novel cereals with increased value in the longer term (Henry 1995). In the sections that follow, the main benefits of genetic engineering are summarised under the following headings:

- productivity
- product quality
- safety.

5.2   Weed control (productivity, quality, safety)

The control of weeds in cereal crops may have a major influence on grain yields but usually has a much lesser impact on grain quality (Kettlewell 1996). This is because the major effect is in early crop growth impacting more on grain number than size. Late weeds are an exception. Generally, contamination of seed crops with weed seeds is likely to be the major quality defect.

5.2.1   Classes of herbicides and available resistance genes
Resistance genes are available for many different classes of herbicide. The major groups of herbicides and the genes available for conferring resistance to these herbicides are listed in Table 5.1.

The most attractive herbicide resistance genes for introduction into cereals are those that confer resistance to herbicides that are considered safe in the environment. Herbicides with low mammalian toxicity and little or no other environmental problem may be attractive alternatives to the more specific herbicides currently in use. The development of transgenic cereals with resistance to appropriate herbicides may facilitate the reduction in use of less desirable herbicides in agriculture and food production.

5.2.2   Problems of escape of herbicide genes to weeds
A major risk associated with the production of transgenic cereals with herbicide resistance is the possibility that new weeds may result either from escape of the

Table 5.1   Herbicide resistance genes (Henry 1997)

Herbicide	Mode of action
Glyphosate	Inhibits 5-enolpyruvyl shikimate-3-phosphate (EPSP)[1]
Sulphonylureas	Inhibits acetolactate synthesis (ALS)[2]
Imidazolinones	Inhibits acetolactate synthesis
Triazalopyrimidines	Inhibits acetolactate synthesis
2-dichlorophenoxyacetic acid	Auxin action
Phosphinothricin	Inhibits glutamine synthesis
Atrazine	Inhibits electron transport in photosystem II

[1] Prevents the synthesis of aromatic amino acids.
[2] Prevents the synthesis of leucine, isoleucine and valine.

gene into other plants or by the transgenic cereals themselves becoming weeds. The production of transgenic plants with resistance to herbicides that have unique modes of action is highly desirable. Multiple herbicide resistance may arise if genes target biochemical pathways that are associated with the action of several classes of herbicide. The problem of escape of herbicide resistance genes from cereals is likely to be a more serious issue for species that are out-crossing rather than for those that are predominantly or exclusively self-pollinating. In some species such as rice, weedy varieties have developed as a result of current agricultural practices. The introduction of herbicide-resistant rices could lead to the development of a further class of weedy rices based upon their herbicide resistance if appropriate agricultural practices are not adopted in association with the new varieties. This could require production to be limited to specific regions and to include the rotation of herbicides or varieties. Despite these limitations, herbicide-resistant cereals should provide enormous advantages in the enhancement of cereal productivity.

5.3   Disease resistance (productivity, quality, safety)

5.3.1   Disease resistance

Diseases may seriously reduce grain quality. Fungal diseases may have a large impact on grain quality (especially grain size) because they may be active on the leaves during grain filling or directly infect the head. Insect pests may reduce yield and grain quality. Post-harvest damage from insects can be a major problem (Mills 1996). Mycotoxins resulting from fungal growth on the grain are a serious safety issue for grain from some environments.

Development of pest- and disease-resistant cereals provides a major opportunity for enhancing cereal productivity. In many environments single diseases may be associated with very serious losses in grain yield. Breeding resistant varieties has been a major strategy used in increasing cereal yields. Transgenic cereals with high levels of disease resistance may extend the options available from conventional plant improvement. Resistance to a wide range of biotic factors may be engineered using appropriate genes. Resistance to viruses, bacteria, fungi, nematodes and insects has been reported.

5.3.2   Manipulation of expression of native genes for disease resistance

Classical cereal breeding has involved the combination of disease-resistant genes from different sources to produce commercial varieties with effective disease resistance (Hammondkosack and Jones 1997). Molecular markers or direct analysis for the presence of the required gene are now used to improve the efficiency of selection of disease-resistant lines in breeding.

One option for the engineering of cereals with disease resistance is to manipulate or enhance the levels of expression of genes already present in the genome. Specific options include the expression of defence genes using constitutive promoters such that the defence gene product was always produced by the plant regardless of the presence of a specific pathogen. Promoters induced by the disease are also an important option. This approach allows defence gene products to be produced by the plant only in response to attack by the pest.

5.3.3   Novel genes for disease resistance

Novel genes from other plants or non-plant sources (Bowles 1990) may provide durable resistant genes for use in cereals (Table 5.2). Examples have been described for bacteria, fungi, nematodes and insects (Shewry and Lucas 1997). The use of virus-derived sequences for breeding virus-resistant plants has been a notable success in many species (Buck 1991, Malik 1999).

5.3.4   Environmental impact of disease resistance

Transgenic plants with resistance to pests and diseases may have a significant impact on the environment (Dale and Irwin 1998). The risks should be similar to

Table 5.2 Some novel pest- and disease-resistant genes of potential value in cereals (Malik 1999)

Type of protection	Gene	Source
Virus	Coat protection	Barley yellow dwarf
	Coat protection	Maize chlorotic mottle Maize chlorotic dwarf Maize chlorotic mosaic
Bacterial and fungi	Chitinase Glucanase	Various Various
Insects	Bt toxin	*Bacillus thuringiensis*

those of traditional resistance breeding. However, if the transgenic strategies prove dramatically more effective they could seriously deplete populations of plant pathogens and even lead to the extinction of highly specific pest organisms. Evaluation of these risks is important in successful applications of transgenic technology (Bergelson *et al.* 1999).

5.4   Improved nutritional properties (quality, safety)

5.4.1   Importance of cereals in human and animal nutrition

Cereals are a very important part of human diets. The three major species, wheat, maize and rice, account for a large proportion of the calories and protein in human diets. The importance of cereals in the food chain is also attributable to the extensive use of cereals in the diets of animals. The major constituents of cereals are the carbohydrates and proteins. Other grain components such as lipids and vitamins may be of great significance in human nutrition because of the large contribution of cereals to the diet. Biotechnology provides new options for manipulation of the nutritional properties of cereal grains. The carbohydrates of cereals include the simple sugars, the more complex oligosaccharides such as fructans, storage polysaccharides of the grain (starch) and the cell wall polysaccharides, all of which are of nutritional value. All of these carbohydrate components are potential targets for manipulation in improvement of cereal quality. (For example, sugar beet has been transformed to produce fructans (Sevenier *et al.* 1998).) Benefits that may result include reduced cariogenic bacteria (dental health), lower energy value and stimulation of beneficial bacteria in the colon. The sugar content may also influence the quality of the grain for various products. Fructans may be considered to be important to human nutrition because of their possible role as soluble fibre (Ninees 1999). Starch, as the major component by weight of the grain, may have a great impact on nutritional quality. Resistant starches (not digested in the gut) may be considered critical in influencing the incidence of certain human diseases, such as heart disease. The cell wall polysaccharides may also be important as either soluble or insoluble fibre, depending on the composition of the polysaccharides in the cereal product. Soluble fibres may reduce the risk of heart disease while insoluble fibres contribute to reduced risk of colonic cancers.

Cereal proteins are not well balanced in amino acids required in a nutritionally balanced diet, and genetic engineering may provide opportunities to improve the balance of essential amino acids in cereal-based diets. The lipids in cereals are generally of limited importance in human nutrition but may be important in animal diets. The manipulation of iron levels in cereals through the introduction of haemoglobin illustrates the potential application of biotechnology to enhancing the nutritional value of cereals.

5.4.2   Animal foods

The major requirements of animals differ depending on whether they are monogastric or ruminant animals and the potential of biotechnology to improve the nutritional values for these two classes of animals differs. Ruminants have a much greater capacity to digest cereal fibre effectively. Soluble fibre components may also be important. For example, the β-glucans of barley limits the use of this cereal in the diet of chickens because of its adverse impact on nutrient utilisation.

5.4.3   Aquaculture foods

The increasing shortage of seafoods (declining fish stocks in the oceans and increasing human population) indicates great potential for enhanced use of cereals in aquaculture diets used in fish farming (Sarac and Henry 1998). Cereals provide a very cheap option and if biotechnology can be used to enhance the nutritional value of cereals as a component of aquaculture diets we can expect wide-scale use of cereals for the production of aquaculture products. Cereals may have an important role as a binder in aquaculture feeds. Improvement of protein and lipid composition by genetic engineering may produce more useful cereal aquaculture feeds.

5.5   Improved processing properties (productivity, quality, safety)

The quality requirement of cereal processors may be complex as indicated for barley in Table 5.3. This table lists a few of the characteristics defined as requirements of a barley for use in malting and brewing. Establishing

Table 5.3   Barley quality characteristics required for brewing (Henry 1990)

Character	Requirement
Grain colour	Bright, white aleurone
Grain size	Plump grains, 90% above 2.5 mm, as little as possible below 2.2 mm
Protein content	Optimum 10.5%–11.5% (dry basis)
β-glucan content	Low – maximum and minimum not defined
Husk	Minimum required for brewing
Dormancy	As little as possible (but no pre-harvest sprouting)
Rate of modification	As fast as possible (steeping and germination maximum 110h)
Malt extract	As high as possible (only too high if protein on husk becomes limiting)
Malt enzymes	As high as possible (minimum requirements for diastatic power and β-glucanase)

opportunities for quality improvement requires a knowledge of the processes used to convert cereals into end products.

5.5.1 Introduction to the range of processes that might be manipulated genetically

The processing of cereals involves a wide range of techniques with very differing raw material requirements. Several of the major processes of cereal processing will be described here in an attempt to identify the major opportunities for biotechnology to be applied to improving cereal processing quality.

Milling

The milling of cereals (Fig. 5.1) involves processes that are dependent substantially on the anatomical structure of the grain. The potential for single genes to be manipulated in ways that enhance milling quality may be limited because of the large number of grain characteristics contributing to milling performance. The shape of the grain and adherance of the various outer layers are of great importance. Hardness is also a key attribute in milling. Grain colour is a key quality attribute that is influenced by the milling process. The levels of pigments and the size of particles generated in milling both influence colour. The contamination of endosperm fractions with more highly coloured outer layers such as bran are key determinants of colour (Ziegler and Greer 1971).

Baking

Many components of cereals such as wheat are important in baking quality and are obvious candidates for the application of biotechnology. The requirements of modern high-speed plant bakeries are for a very consistent quality. Characteristics such as dough stickiness associated with some alien (non-wheat) sources of disease resistance in wheat are serious problems in these high-capacity facilities.

The major components of wheat all contribute to baking quality and are thus targets for genetic improvement of baking quality. Proteins are essential for the viso-elastic properties of wheat doughs. Starch and cell wall polysaccharides (e.g., pentosans) also influence baking quality. The breakdown of starch by amylases is a key process in baking. The pentosans of the cell wall also have a significant influence on loaf quality (Pomeranz 1971).

Malting

Malting is the first step of processing grain for use in brewing and distillation. The malting of cereals, most specifically barley, is a process of germination. The rate of germination and the changes in the composition of the barley during malting are potentially important targets for biotechnology. The breeding of malting barley varieties focuses on several major quality characteristics associated with the malting process.

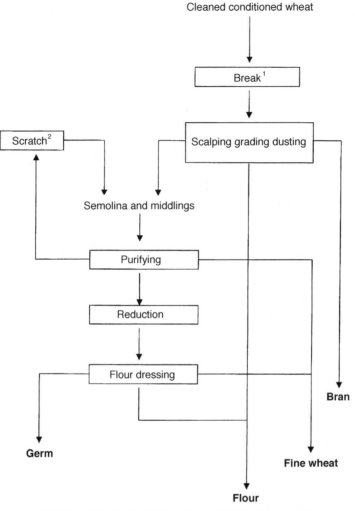

Cleaned conditioned wheat

Break[1]

Scratch[2]

Scalping grading dusting

Semolina and middlings

Purifying

Reduction

Flour dressing

Bran

Germ

Fine wheat

Flour

1 · This process opens the grain to allow the endosperm to be scraped from the bran.
2 · Separates bran and endosperm.

Fig. 5.1   Schematic of flour milling (based on Kent and Evers 1994).

Enzymes are involved in the breakdown of cell walls during malting. The β-glucanases expressed during germination are essential to ensure that the levels of β-glucan in the malt are low. High malt β-glucan levels contribute to high wort viscosity, poor wort and beer filtration and potential hazes in finished beer.

Brewing
The brewing of beer from malt requires specific malt specifications that are potentially able to be manipulated using biotechnology. Diversification of beer styles and markets are imposing divergent raw material requirements for the different beer styles.

Sufficient levels of starch degrading enzymes in the malt are necessary to ensure breakdown of starch to fermentable sugars during brewing. The amount of starch (adjunct) added in the form of rice or maize is a key determinant of the level of starch degrading enzyme required in the malt. The relative levels of different starch degrading enzymes are important in determining the nature of the substrate for fermentation and the sugar, alcohol and oligosaccharide content of the beer. The alcohol and residual sugar content (sweetness) are influenced by the levels of fermentable sugars and the non-fermentable oligosaccharides contribute to the taste (mouthfeel).

Distilling
Distillation is a process where the product quality may be less dependent on the raw materials than many other processes and may be a less important target for biotechnology application in relation to the cereal raw material. However, the quantity of starch available for fermentation might be enchanced.

Extrusion
Extrusion (production using high temperatures and pressures) of cereal products is an increasingly important process in the production of a wide range of products including snack foods, breakfast cereals and pet foods. The processing properties required are complex but may be enhanced by the application of biotechnology.

5.5.2 Wheat utilisation
Wheat is used for a wide range of products with differing quality requirements (Fig. 5.2). The importance of different wheat grain components and characteristics depends on the ultimate end use product (Morris and Rose 1996). The protein quality is much more important for products such as breads than for cakes and biscuits. Enhanced levels of desirable high molecular weight glutenins may be desirable in wheat for use in breadmaking. Manipulation of starch synthesis and starch properties may be more important in products such as noodles. Colour may be controlled by a small number of genes and has differing importance. For example, the yellow pigments in durum wheat are considered highly desirable while some noodle products require very white flour. The improvement of specific attributes using biotechnology needs to target characters specific for particular end uses. Wheat is used to produce the types of food products listed in Table 5.4 (Morris and Rose 1996).

Improvement of the value of wheat for this wide diversity of uses requires the matching of wheat characteristics to specific end product requirements. For example, wheat with different combinations of protein and hardness are better suited to particular end uses (Fig. 5.2). However, some combinations of characteristics will not be optimal for any major end use.

Genetic improvement of wheat quality needs to address targets relevant to the end use characteristics of wheat from a particular environment or region. For

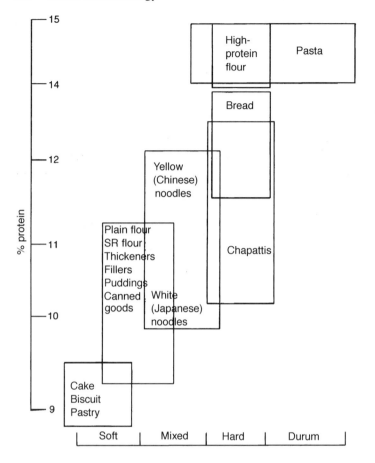

Fig. 5.2   The broad range of wheat types required for different uses (Moss 1978).

example, selection for specific starch metabolism mutants or engineering of improved starch qualities (such as starch-pasting properties) may be important in regions producing noodle wheats while storage protein modification may be more appropriate in regions producing bread wheats (Anderson 1996).

5.5.3   Beer production

The production of a specific product such as beer can be the basis for analysis of opportunities for application of biotechnology. The traditional process from barley to beer is depicted in Fig. 5.3. Some key attributes for malting and brewing have been described above. The efficiency of this process may be influenced by the composition of raw materials (especially the cereals) used and the type of beer to be produced.

Changing composition of barley by genetic engineering might alter performance in any of the many steps in the process of beer production (Henry

Table 5.4   Food products produced from wheat

Fermented (leavened) breads	white pan breads	white pan loaves, sandwich (hamburger buns), raisin breads, variety breads
	hearth breads	baguettes, Vienna breads, sourdough breads
	sweet goods	Doughnuts, cinnamon rolls, coffee cakes, danish, puff pastries, French brioche
	steambreads	Chinese northern style steambread, Chinese southern style steambread, *Pan de sol* (Philippines), *Saka-manju* (Japan), *Mushi-manju* (Japan)
Flat breads and crackers	flatbreads	chapatti, rotti, naan, paratha, poori, balady, pia, banabri
	tortillas	
	pizza crust	
	muffins	
	crumpets	
	bagels	
	pretzels	
	crackers	soda crackers, cream crackers, water biscuits, graham crackers, sprayed crackers, savoury crackers
Cookies and cakes	pie crusts	
	cookies (biscuits)	
	scones	
	moon cake	
	batters (ice cream cones, pancake and waffles)	
	cakes	
	tempura	
	soup thickeners	
Noodles	alkaline noodles	
	white salted (Udon)	
	soba	
	buckwheat	
	egg	
	noodles may be fresh, dried, frozen or instant.	
Breakfast foods Starch/gluten Pasta and durum products	pasta	spaghetti, macaroni, lasagne, fettuccine
	couscous	
	bulgur	
	durum bread	

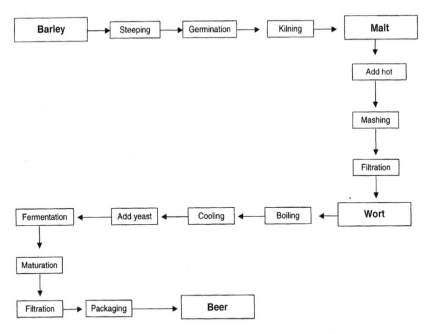

Fig. 5.3   Traditional beer brewing (Henry 1996).

1996, McElroy and Jacobsen 1995). The production of proanthocyanidin-free barley to reduce haze and improve the shelf life of beer in cold storage has been achieved using mutants and may be more precisely controlled using genetic transformation to block specific steps in the pathways leading to proanthocyanidin formation. One complication of this change may be that the absence of proanthocyanidins in the wort may result in reduced protein precipitation during boiling. The overall result can be a decline in product stability due to increased levels of protein in the beer. This illustrates the need to understand all the interactions during processing before implementing changes to novel cereal raw grains.

Molecular markers have been developed for many malting quality attributes and these may accelerate the development of new malting quality barley varieties for use in beer production (Han *et al.* 1997).

Genetic engineering of barley to improve cell wall breakdown during malting and brewing (Fincher 1994) and lipids contributing to beer flavour (Hoekstra *et al.* 1994) have been the subject of active research. Modifications of starch properties and metabolism and protein composition may also be important (Edney 1996). A fundamental limitation is that extract levels are limited by the need to preserve a minimum amount of husk to act as a filter bed in many traditional brewing processes. Grains with extremely high proportions of endosperm necessary for extreme extract levels will have insufficient husk. New filtration technologies could overcome this limitation allowing the development of very high extract barleys. Reliance on enzymes in the malt for starch breakdown during mashing and the issue of yeast nutrition also requires that a minimal protein level be maintained. Genetic engineering could allow a higher proportion of proteins in the malt either to contribute useful enzyme activity or support yeast nitrogen nutrition.

5.6   Improved cereal quality control (quality, safety)

The value of cereals can be improved by better specification of the identity and composition of cereals in marketing. The major contributions to achieving this type of value addition are quality assurance programs and enhanced tools for analysis of products for chemical and microbial contamination and for genetic identity and purity.

5.6.1   Chemical and microbial purity
The strict application of quality assurance principles can ensure cereal safety and value. New technologies allow the use of ELISA tests to establish the level of pesticide residues and DNA-based analysis of microbial contaminants.

5.6.2   Genetic purity
The identity of a cereal genotype defines many of the quality characteristics such as protein content and grain size. This can define the processing value and optimal end use of a parcel of cereal grain. Protein content can currently be measured rapidly in the field using near infra-red reflectance (NIR). The development of rapid genotyping methods would find wide application in the cereal industry and allow a relatively complete objective description of samples for commercial valuation.

Distinction of high-value noodle wheats from visually similar wheats of low noodle quality, and distinction of malting and food barleys with similar appearance, are good examples of the areas of potential application of this technology. Genetic purity and level of admixture are also important attributes to assess in commercial trading because of the financial incentive to add low-value grain to parcels of very high value but visually similar genotypes. Rapid DNA extraction from grain is a key technical requirement for successful application of these methods.

5.7   Summary: future prospects and limitations

Biotechnology is likely to have a major impact on the value of cereal production both by increasing productivity and by improvements in product quality. The improved productivity is likely to result initially from the removal of biotic stress constraints associated with major pests and diseases. Herbicide resistance is an option that is likely to be able to achieve early adoption and success. Improvements in grain quality are likely to be generally more difficult to achieve. The major attraction of biotechnology is the possibility of introducing totally new or novel characteristics into cereals that will result in products with characteristics outside the range of those currently available. A major limitation to the introduction of such characteristics is the requirement of cereals to be compatible with existing processes of cereal food production. Market resistance to products requiring new processing techniques will come from the large investment that may be required to develop new processing facilities. Improved methods of quality control and analysis of product identity and purity will enhance the value of cereal products. Adverse consumer attitudes are also a significant risk if transgenic products are not well designed and marketed.

5.8   Sources of further information and advice

Key books
The following books are sources of further information:
Cereal grain quality. R J Henry and P S Kettlewell (eds) Chapman and Hall, London 1996.
Principles of cereal science and technology. Carl Hoseney. American Association of Cereal Chemists, St Paul Minnesota 1994 (2nd edn).
Practical applications of plant molecular biology. R J Henry. Chapman and Hall, London 1997.
Kent's Technology of Cereals (4th edn) N L Kent and A D Evers. Woodhead, Cambridge 1994.
Improvement of cereal quality by genetic engineering. R J Henry and J A Ronalds (eds) Plenum, New York 1994.
Agri-Food Quality. An interdisciplinary approach. G R Fenwick, C Headly, R L Richards and S Khokhar (eds) The Royal Society of Chemistry, Cambridge 1996.
Applied Plant Biotechnology. V L Chopra, V S Malik and S R Bhat (eds) Science Publishers USA 1999.
Alternative End Uses of Barley. D H B Sparrow, R C M Lance and R J Henry (eds) Royal Australian Chemical Institute, Melbourne 1988.

Major trade/professional bodies
American Association of Cereal Chemists. St Paul, Minnesota.
International Association for Cereal Science and Technology (ICC). Vienna.
Royal Australian Chemical Institute Cereal Chemistry Division. Melbourne.
The Institute of Brewing. London.
International Society for Plant Molecular Biology. Athens, Georgia.
Association of Applied Biologists. Wellesbourne, UK.
Plant and Animal Genome conferences – San Diego, California.

5.9   References

ANDERSON O Molecular approaches to cereal quality improvement in *Cereal Grain Quality* Chapman and Hall, London (1996) pp 371–404.

BERGELSON J, WINTENER J and PURRINGTON C B Ecological impacts of transgenic crops in *Applied Plant Biotechnology* Chopra V C, Malik V S and Bhat S R (eds). Science Publishers Inc, USA (1999) pp 325–43.

BOWLES D J Defence-related proteins in higher plants. *Ann. Rev. Biochem* (1990) **59** 873–907.

BUCK K W Virus resistant plants in *Plant Genetic Engineering* Grierson D (ed.) Blackie, USA (1991) pp 131–78.

DALE P J and IRWIN J A Environmental impact of transgenic plants in *Transgenic Plant Research* Harwood Academic Publishers, Amsterdam (1998) pp 227–85.

EDNEY M Barley in *Cereal Grain Quality* Henry R J and Kettlewell P S (eds) Chapman and Hall, London (1996) pp 113–51.

FINCHER G B Potential for the improvement of malting quality of barley by genetic engineering in *Improvement of Cereal Quality by Genetic Engineering* Henry R J and Ronalds J A (eds) Plenum Press, New York (1994) pp 135–8.

HAMMONDKOSACK K E and JONES J D G Plant disease resistance genes. *Ann. Rev. of Plant Physiology and Plant Molecular Biology* (1997) **48** 575–607.

HAN F, ROMAGOSA I, ULLRICH S E, JONES B L, HAYES P M and WESENBERG D M Molecular marker-assisted selection formulating quality traits in barley. *Molecular Breeding* (1997) **3** 427–57.

HENRY R J Barley quality: an Australian perspective. *Aspects of Applied Biology* (1990) **25** 5–14.

HENRY R J Biotechnology applications in the cereal industry: Procedures and prospects. *Cereal Foods World* (1995) **40** 370–3.

HENRY R J Improvement of the quality of barley for beer production in *Agri-food Quality – an interdisciplinary approach* Fenwick G R, Hedley C, Richard R L and Khokhar S (eds). The Royal Society of Chemistry, London (1996) pp 15–18.

HENRY R J *Practical applications of plant molecular biology.* Chapman and Hall, London (1997).

HENRY R J and KETTLEWELL P S (eds) *Cereal Grain Quality.* Chapman and Hall, London (1996).

HENRY R J, KO H L and WEINING S Identification of cereals using DNA-based technology. *Cereal Foods World* (1997) **42** 26–9.

HOEKSTRA S, VAN ZIJDERVELD M, VAN BERGEN S, VANDER MARK E and HEIDEKAMP F Genetic modification of barley for end use quality of barley by genetic engineering in *Improvement of Cereal Quality by Genetic Engineering* Henry R J and Ronalds J A (eds). Plenum Press, New York (1994) pp 139–44.

KENT N L and EVERS A D *Technology of cereals*, 4th edn Woodhead Publishing, Cambridge (1994).

KETTLEWELL P S Agronomy and cereal quality in *Cereal Grain Quality* Henry R J and Kettlewell P S (eds). Chapman and Hall, London (1996) pp 407–37.

McELROY D and JACOBSEN T What's brewing in barley biotechnology? *Bio/Technology* (1995) **13** 245–9.

MALIK V K Biotechnology: Multi-billion dollar industry in *Applied Plant Biotechnology* Chopra V L, Malik V S and Bhat S R (eds). Science Publishers, USA (1999) pp 1–69.

MILLS J T (1996) Quality of stored cereals in *Cereal Grain Quality* Henry R J and Kettlewell P S (eds). Chapman and Hall, London (1996) pp 441–78.

MORRIS C F and ROSE S P Wheat in *Cereal Grain Quality* Henry R J and Kettlewell P S (eds), Chapman and Hall, London (1996) pp 1–54.

MOSS H J Factors determining the optimum hardness of wheat. *Aust J Agric Res*. (1978) **29** 1117–26.

NINEES K Breakfast foods and the health benefits of inulin and oligofructose. *Cereal Foods World* (1999) **44** 79–81.

POMERANZ Y Composition and Functionality of Wheat flour components in *Wheat Chemistry and Technology Volume III* Pomeranz Y (ed.) AACC St. Paul (1971) pp 585–674.

SARAC H Z and HENRY R J Use of cereals in aquaculture production systems in *Pacific People and Their Food*, American Association of Cereal Chemists (1998) pp 193–217.

SEVENIER R, HALL R D, VAN DERMEE I M, HAKKERT H J C, VAN TUNEN A J and KOOPS A J High level fructan accumulation in a transgenic sugar beet *Biotechnology* (1998) **16** 843–6.

SHEWRY P R and LUCAS J A Plant proteins that confer resistance to pests and diseases, *Advances in Botanical Research 26* Academic Press (1997) pp 135–92.

ZIEGLER E and GREER E N Principles of milling in *Wheat Chemistry and Technology Volume III* Pomeranz Y (ed.) AACC St. Paul (1971) pp 115–99.

6

Molecular biological tools in cereal breeding

W. Thomas, Scottish Crop Research Institute, Dundee

6.1 Introduction

6.1.1 Plant breeding

The small-grained cereals considered in this chapter, namely wheat, barley and oats, are all natural inbreeders, i.e. any one plant is more likely to reproduce with itself rather than with a neighbour. This means that, once a potential plant variety is homozygous, there is no residual variation and it will reproduce itself exactly from one generation to another. Breeding programmes for the small-grained cereals generally follow a pedigree selection scheme with minor variations in detail. At the beginning, breeders cross parents which complement each other for desirable characteristics to produce the F1 generation which will be genetically uniform. The F1 generation naturally self-pollinates to eventually give rise to a population of inbred lines that will contain 2^n different genotypes, where n is the number of loci at which the two parents differ. With differences at just 20 loci, the number of potential different inbred lines is 3^{20} or 1,048,576 – a large number that would occupy about a quarter of a hectare at UK commercial sowing rates if each line was represented by just one seed. To be sure of generating one particular gene combination, the population size would need to be larger still and when one considers that a breeder will make several hundred crosses a year it is abundantly clear that plant breeding programmes cannot accommodate all the possible combinations. Most breeding programmes are therefore constrained to several hundred thousand F2 plants spread over a range of crosses that a breeder will have judged from parental performance to be the best.

I thank SERAD for funding my core research work, Drs Machray, Meyer and Russell for allowing me to quote unpublished work and Drs Ellis, Forster and Swanston for their helpful comments.

Breeders will select the best F2 plants on the basis of easily recognised characters such as disease resistance, height and maturity. Further selection on such characters is practised upon the F3 generation, after which some breeders will then carry out a preliminary yield trial at the F4 generation and suitability for processing such as baking wheat and malting barley will also be assessed. Yield and processing characters are affected to varying degrees by the environment in which the crops are grown so that the ranking of individuals may vary between environments. This is known as genotype × environment interaction and, to get a more complete picture of the potential of breeding lines, further yield trials are carried out at a number of sites in one or more subsequent generations. At the end of this series of trials, breeders will have assembled a comprehensive body of data upon their remaining selections and will enter the best into official trials. In the UK, approximately 120 winter wheat, winter barley and spring barley lines are entered into National List trials each year in total. National List trials take two years and are conducted according to set protocols in order to select the best three to ten from each crop to enter Recommended List Trials, which are carried out at a number of sites all over the UK. After a year, sufficient data will have been gathered upon the entries to enable a provisional recommendation to be given to between one and five new varieties from each crop each year, although not even one may be widely grown for more than one or two years.

From the above brief outline of a conventional cereal breeding scheme, it can be deduced that such schemes are lengthy, occupying up to ten years to provisional recommendation for the small-grained cereals in the UK and twelve years to make a commercial impact. Bingham and Lupton[1] provide a comprehensive review of a practical winter wheat breeding programme, which is typical of many other cereal breeding programmes. Given the huge investment required to run such a scheme over a 12-year cycle, and the competition provided by others, breeders are keen to exploit alternatives which will either shorten the time-scale or increase the efficiency of identification of elite lines and especially any which combine both elements. The developments in molecular biological tools over the past 15 years have led to the possibility of direct selection for the genetic constitution, or genotype, of individuals. This is an attractive alternative as it reduces the amount of selection upon character measurement, or phenotype, when the environment and its interaction with genotype can modify the expression of genes. Conventional plant breeding schemes have been outstandingly successful, contributing an average 1% yearly increase in the yield of barley and wheat.[2] The deployment of molecular biological tools in cereal breeding must either lead to more efficient ways of achieving this rate of progress or lead to increased breeding progress.

6.1.2   Phenotype

The phenotype of a line is the result of the interaction of its genes with the environment. Thus a character with a high degree of genetic control is less affected by varying environmental conditions than one with a low degree of genetic control. Phenotypic selection for performance is reliable in the former case but is likely to be unreliable in the latter. Breeders can therefore make selections for the former group of characters in unreplicated trials or nurseries in the early stages of a programme and reduce their populations to more manageable levels for field trials. Examples of characters that breeders select for in early generations are disease resistance, maturity and short straw. Many of these characters are controlled by a single gene of large effect (major gene) and are therefore qualitative characters where phenotypic expression is a good indicator of a line's genetic constitution. Most important characters such as yield and quality are, however, controlled by a number of genes, each often of small effect. In such cases, because variation is generally continuous and it is impossible to discriminate between the different genotypes the characters are termed quantitative. Each individual gene often has a small effect, is subject to considerable environmental modification, and genotype × environment interactions may occur. Accurate prediction of such characters from a single unreplicated trial is extremely unreliable and therefore breeders carry out multi-site trials over two or more seasons to get an estimate of a line's phenotypic potential for characters such as yield and quality.

6.1.3   Genotype

The genotype defines the exact genetic constitution of a line but this is unknown for most characters, as the actions of minor genes do not individually produce discernible effects. Indeed, the action of many major genes, such as major gene disease resistance, cannot be recognised without extensive testing with race-specific isolates. This means that breeders are usually completely reliant upon phenotype as a predictor of a line's performance. Knowledge of a line's genotype at individual loci, however, can be used to build up a picture of the relationship between loci. Loci that tend to segregate together are said to be linked and the closer the loci are, the greater the degree of linkage and the fewer the recombinants. Knowledge of such relationships is of value to plant breeders as they can then predict which parental combinations will be most likely to give desirable recombinants. Furthermore, if a breeder wished to find recombinants between particular loci, knowledge of linkage relationships can be used to formulate population sizes necessary to produce as many of such lines as desired. Assembling segregation data on a number of loci enables construction of genetic linkage maps, which can be used in conjunction with phenotypic data to reveal regions of the genome that control a character. These regions are termed quantitative trait loci (QTL) and can also be placed on genetic maps.

6.2   Markers

6.2.1   Morphological, isozyme and protein

Major gene variation has been known for a long time with many morphological mutants being described. These were used to construct early genetic maps for barley and permitted the addition of major disease resistance genes as they became known. However, many morphological mutants are deleterious and were of little benefit in identifying regions of the genome controlling economically important characters. The effect of some major genes upon other traits could be analysed in populations segregating for the gene. Such studies were used to identify deleterious effects upon yield and quality characters at two dwarfing loci used in spring barley.[3–6] The deployment of *Hordeum laevigatum* as a source of disease resistance has also led to deleterious associations, probably through introgression of undesirable alleles together with the disease resistance genes.[7] However, such findings became known after the genes had been widely deployed and served to explain observed results rather than advocate the selection of alternative genes. Isozyme markers have also been proposed to augment selection and a number of loci have been included on classical genetic maps of barley[8] and wheat.[9] Some isozymes have been used to select for specific genes, e.g. the endopeptidase gene on wheat chromosome 7D to select for the VPM1 eyespot resistance gene.[10] Analysis of the high molecular weight glutenin sub-units was found to be a good predictor of bread-making quality of wheat and such analyses are now carried out routinely in wheat breeding programmes prior to baking tests, thus increasing the efficiency of selection.[11] In general, the variation at isozyme loci is insufficient to be of great practical use.

6.2.2   Molecular markers

Maps constructed from morphological, isozyme and disease resistance genes were the composite picture developed from the integration of a large number of small-scale experiments involving three or four linked loci. The development of methods to assay DNA sequence differences, or polymorphism, permitted the construction of a map of the whole genome from one single population. This has a number of advantages as a much better estimate of the overall order of loci along the chromosomes can be obtained.

Restriction fragment length polymorphism (RFLP)
RFLPs rely on the use of restriction enzymes to digest genomic DNA where polymorphism can arise through mutations to create or remove restriction sites or by sequence deletions or insertions between sites. After digestion with restriction enzymes, the DNA fragments are separated by electrophoresis and labelled DNA probes that recognise specific sequences are then hybridised to the fragments and polymorphisms are detected as length differences.[12] Because different RFLP alleles are recognised on the same gel, the marker system is said to be co-dominant as heterozygotes (or mixtures) produced both allelic bands.

The initial use of RFLPs in the small-grained cereals was hampered, to some extent, by the relatively low levels of polymorphism that they detected in cultivated germplasm, particularly in wheat. Because RFLP analysis is based upon hybridisation assays, comparatively large amounts of DNA of a reasonably high standard of purity are required, which is another limiting factor. A major advantage of RFLPs, however, is the fact that they provide robust anchor loci that are easily distinguished not only across different crosses within a species but also can be used as inter-specific anchor loci. This latter feature has been of great value in comparative mapping within the *Triticeae*, leading to the ordering of homoeologous loci across a number of species.[13]

Random amplified polymorphic DNA (RAPD)
The creation of mapping data using RFLPs was laborious, as it required hybridisation. The development of molecular marker techniques based on the polymerase chain reaction (PCR) to amplify target DNA segments prior to detection resulted in a quantum leap in the generation of data points. RAPD markers were generated by using random primers in the range of 10–20 nucleotides to detect complementary sites across relatively short distances within the genome.[14] The presence of complementary sites in a genotype resulted in the primer amplifying a DNA fragment under set PCR conditions. A number of different fragments could be amplified with each primer and separated electrophoretically. The absence of a complementary site results in the absence of a DNA fragment to reveal a polymorphism. This type of marker system is dominant, as a band is either produced or not produced, and the inability to distinguish heterozygotes from a homozygous class is a major disadvantage of the system. Whilst RAPDs could generate data much more quickly than RFLP, they suffered additional disadvantages through difficulties in repeatability, while bands from one cross could not readily be transferred to other crosses. The development of new marker types soon rendered RAPDs obsolete.

Amplified fragment length polymorphism (AFLP)
AFLPs are generated from restriction digests of genomic DNA with enzymes that recognise both frequent and rare sites. This generates a large number of fragments, to which adapters of known sequences are joined (ligated), and then multiplied for several PCR cycles using non-selective primers (pre-amplification). Selective PCR amplification is then carried out with primers that recognise the adapter sequence plus 1–3 random bases. The amplified fragments are then separated electrophoretically and recognised through the use of radioactive or fluorescent labelling.[15] Like RAPDs, polymorphism is established through the absence of a fragment, which results in an essentially dominant marker system. The more random bases, the fewer the bands. There is therefore a trade-off between the ease of recognising unique bands on the gel and the amount of data generated. Nevertheless, each primer combination generates a large number of bands, 10–20 of which may, on average, be polymorphic.[16] AFLPs are a much

more robust and versatile marker system than RAPDs and their ability to generate a large number of polymorphic bands for each primer combination gives the system a high multiplex ratio.[17] The use of AFLPs therefore generates a large number of data points in a short time and greatly accelerated the development of genetic maps. Waugh *et al.*[18] compared AFLP loci in three barley crosses and found that they mapped to the same chromosomal locations, so AFLP markers could be used as anchor loci within a species. Precise sizing of AFLP fragments was, however, required to establish identical loci across several crosses and this was not always possible or practical, limiting the usefulness of the system in comparative mapping.

Simple sequence repeats (SSRs)
SSRs, or microsatellites, are short tandem repeats of mono-, di-, tri- and tetra-nucleotides although more complex repeats have been detected. They are abundant in mammals[19] and have been found in plants by searching sequence databases.[20] They can also be found by sequencing random clones of genomic DNA. The main advantage of SSRs over RAPDs and AFLPs is that they are, like RFLPs, co-dominant markers and can therefore reliably identify heterozygotes. They have also proved to be multi-alleleic, with over 30 different alleles being detected by a single SSR primer pair in barley (J.R. Russell, pers. comm.). This facet renders them highly informative so they have high polymorphic information content (PIC)[21] values. The main drawback of SSRs is their high development cost, requiring extensive DNA cloning and sequencing for their identification. The production of libraries enriched for specific SSR motifs has been found to greatly increase the rate of SSR discovery[22] but the process is still resource hungry. A number of groups have invested resources in developing SSRs and 230 are available for wheat[23] and over 560 for barley.[22] Due to their high PIC values, SSRs are proving to be excellent anchor loci for comparing various studies and are the current marker system of choice within a species. Primers for SSR loci developed for one species, however, usually fail to amplify a product in another species and, even if a product is produced, it cannot be relied upon as a homoeologous locus.

Derivative marker systems
A number of marker systems have been developed that share features of the other PCR based systems but differ in a number of respects. Two methods based on simple sequence repeats are inter-simple sequence repeats (ISSR)[24] and *Copia*-SSR,[25] also termed retrotransposon-microsatellite amplified polymorphism (REMAP).[26] Two other methods, which are based on retrotransposons, are sequence-specific amplified polymorphism (S-SAP)[27] and inter-retrotransposon amplified polymorphism (IRAP).[26] All four are PCR based and produce a number of polymorphic bands for each PCR reaction. Results published so far suggest that all have a higher multiplex ratio than AFLPs, e.g. ISSRs in rice,[28] S-SAPs in barley,[27] IRAP and REMAP in barley.[26] All are dominant marker systems and are therefore less informative than SSRs and their use as anchor loci

has not been evaluated, but they do seem to have some promise as an alternative to AFLPs.

Single nucleotide polymorphisms (SNPs)
SNPs occur as a result of a single base substitution, deletion or addition in a sequence and have been detected at a rate as high as 1 in 26 base pairs in barley (R.C. Meyer, pers. comm.). The advantage of SNPs is that specific oligonucleotides can be designed to detect a particular polymorphism in a positive or negative fashion. One can then analyse a batch of material with a high throughput and automated assay. Such a system also offers the possibility of dispensing with gel-based detection methods, which are one of the current limiting factors in marker analysis. As more sequence data become available, particularly from the commitment of the public and private research communities to develop large libraries of clones representing the expression of single genes (expressed sequence tags – ESTs), it is becoming more realistic to identify SNPs on a large scale in the small-grained cereals. At present, there is no simple, robust way of detecting SNPs other than amplifying a target sequence for a test panel of genotypes and comparing all the sequences to reveal polymorphisms. This therefore requires a heavy investment in DNA sequencing but is now feasible with the latest generation of sequencers. Although SNPs have been found in barley which are diagnostic for the *rym4* and *rym5* resistance genes to Barley Yellow Mosaic Virus (R.C. Meyer, pers. comm.), it is likely that a combination of three or more successive SNPs in a sequence will be of more value in diagnostics. The genotypic constitution at a number of successive loci (haplotype) gives much more information than the genetic constitution at an individual locus as the possible combinations are much more than the two at a single SNP locus. It is therefore likely that SNPs will become the main marker system in diagnostics but their value in generating genetic maps is unknown as yet, although their deployment would mean that maps would be based on known function rather than largely anonymous markers.

6.2.3 Genetic maps
Genetic maps show the location of genes on chromosomes and the distances between them so that one can estimate whether or not two genes are likely to segregate together or, if breakage of a linkage is desired, the population size necessary to ensure recombinants between two neighbouring loci. Genetic maps are constructed by assembling segregation data for a series of markers, which can be used to estimate the location of genes relative to each other. With morphological markers, segregation could usually only be studied for a few loci at a time. The standard procedure is to compare the frequencies of the various parental and recombinant classes in a segregating population such as an F2 or, more recently, doubled haploids. If two loci are completely unrelated, then the observed frequencies should be within the sampling error of that expected for unlinked loci. If the observed frequency of the recombinant classes is

significantly less than that expected, then the loci are located in the same region and the degree of linkage can be estimated from the observed frequencies of the various marker classes. Thus, placing a new marker gene on a genetic map was a matter of crossing it to a series of other known marker loci and checking the segregation ratios to see if there was any evidence of linkage. This led to the placement of 38 loci across seven linkage groups in barley by 1951[29] and nearly 80 loci were assigned to the seven barley chromosomes by 1962.[30] Methodologies were developed to combine segregation data from a series of separate experiments to produce more inclusive maps[31] which, for barley, were published annually in the *Barley Genetics Newsletter*.

Mapping in this way was very much a matter of trial and error and it was not always possible to locate commercially important major genes. For example, the *Hordeum laevigatum* mildew resistance gene (*MlLa*) was found in the spring barley cultivars Vada and Minerva and many of their derivatives but a comprehensive conventional linkage analysis failed to localise the gene to any one barley chromosome.[32] The development of molecular maps soon enabled the gene to be located on chromosome 2H.[33,34]

Associations of major genes

The assignment of some commercially important major genes to particular chromosomes helped to elucidate some information about the control of quantitative characters. By developing random inbred populations that segregated for an easily characterised major gene, it is possible to assess the whole population for a range of economically important characters. The variation for each character can then be partitioned into that ascribable to differences between the alternative alleles at the major gene locus and the variation within groups. If the former is significant when tested against the latter, then the region of the genome in which the major gene is located also plays a role in the genetic control character measured.[35] This method was used to find that the *sdw1* and *ari-eGP* dwarfing genes in barley were associated with a number of other characters.[3–6] Whilst this analysis gave an indication of the possible consequences of deploying specific major genes, it is essentially retrospective and cannot distinguish whether associations are due to linkage or pleiotropy.

Mapping software

The development of large arrays of molecular marker data meant that manual mapping methodology was far too labour intensive. Computer software was therefore developed to reduce the amount of manual calculation. The principles behind mapping remain the same but were updated to examine a large number of marker combinations simultaneously. The most popular (by citation) mapping software is MAPMAKER,[36] although the more recent JOINMAP[37,38] has gained in popularity, again by citation. GMENDEL[39] has also been used and all three differ in their approach to map construction but essentially produce similar maps from a given set of data. All have a reasonable user interface, are well documented and each has routines to evaluate the quality of the input data and/or the maps produced. They can all cope with the various different types of populations generally used in mapping (F2, backcross (BC), recombinant inbred lines (RIL) and doubled haploid (DH)) inbreeding species such as the small-grained cereals. JOINMAP has the added advantage of being able to develop a composite map from different mapping populations but with some common reference markers.

Physical maps

Genetic maps provide an interpretation of the distribution of genes along chromosomes based purely upon recombination. Using suitable stains, banding patterns can be directly observed on barley chromosomes.[40] Some of these bands differ between barley cultivars and can be scored as genetic markers in segregating populations. Linde-Laursen[41] combined a banding pattern marker with other markers to demonstrate that the genetic map location of the band differed from its physical location. This reflects the tendency for recombination to occur at the ends of cereal chromosomes, rather than at random.[42] The development of fluorescent *in situ* hybridisation (FISH) provides an opportunity to visualise loci directly upon chromosomes and therefore extend the comparison of physical and genetic maps. Laurie *et al.*[43] used ribosomal RNA markers to demonstrate considerable differences between the physical and genetic maps of the long arm of barley chromosome 2H. Kunzel and Korzun[44] found no physical markers in some 70cM of the distal part of the long arm of barley chromosome 3H, indicating that the genetic map was considerably lengthened in comparison to the physical. Differences between genetic and physical maps have important implications for those studying the organisation of the genome but the genetic map represents what it is possible to achieve through recombination in terms of segregation of markers and is therefore the most relevant measure for those studying the inter-relationships of characters.

Published maps

In the early 1990s extensive maps were published for two barley populations, Proctor × Nudinka,[45] and Igri × Franka.[46] Later the North American Barley Genome Mapping project developed extensive maps for two barley crosses, Steptoe × Morex,[47] and Harrington × TR306.[48] The three ancestral genomes of bread wheat necessitated the construction of three sets of maps, with many homoeologous loci that map to two or three of the constituent genomes. The early use of RFLPs in wheat mapping was to construct maps from different crosses for each linkage group (e.g. Chao *et al.*[49]).

The first genome-wide maps of wheat crosses involved its wild relatives[50,51] but, because of the problems of low levels of polymorphism detected by RFLPs, an RFLP-based map of an inter-varietal bread wheat cross[52] took longer to develop. Similarly, the first oat RFLP map to be published was for a cross between wild species[53] with a map from cultivated oats being developed later.[54]

In addition, JOINMAP has been used to construct composite maps of barley over four[55] and seven mapped populations.[56]

RAPDs were mainly used to fill the gaps in maps, using known RFLP loci as points to anchor the data to chromosomes. Several barley maps including RAPD markers were published, e.g. Vogelsanger Gold × Alf[57] and Blenheim × E224/3.[58] Becker et al.[59] first reported the use of AFLPs in barley mapping and Powell et al.[16] and Qi et al.[60] reported other major barley mapping efforts using AFLPs. A largely AFLP-based map of a bread wheat cross has also been developed[61] and AFLPs have also been used in oat mapping.[62] SSRs are now being incorporated into genetic maps of barley[22,63,64] and maps of bread wheat based on SSRs have also been published.[23,65] The most efficient current mapping strategy is to combine a number of reference SSRs with AFLPs and/or the derivative markers noted above to fill in the gaps.

All of the above maps are notable for their relative lack of non-molecular markers. In the barley Proctor × Nudinka map, a major gene mildew resistance at the Mla locus on chromosome 1H and the naked locus n on 7H is presented.[45] The lack of non-molecular markers reflects the lack of major gene morphological variants in crosses between cultivars, be they of barley, oats or wheat. More recently, attempts to incorporate morphological markers on barley molecular maps have been made through the Oregon Wolfe Barleys (http://www.css.orst.edu/barley/wolfebar/wolfnew.htm) and Kleinhofs et al.[66] have attempted to place some barley morphological markers on molecular maps.

With the large and ever-increasing numbers of molecular markers available for mapping in barley, oats and wheat, markers are available that give comprehensive genome coverage of all species. The problem is now establishing an order amongst a huge range of data. It is important to remember that the statistical resolution of genetic distances is dependent upon the number of individuals in the population being measured, as can be seen for the linkage formulae given by Mather.[67] Whilst mapping software usually estimates orders of loci to tenths of a centi-Morgan, these are beyond the resolution of most of the populations being studied and thus the fine-map order of a population of say 100 backcross or doubled haploid individuals cannot be relied upon. To solve this presentational problem, Kleinhofs et al.[66] have developed a system of 'bins' at stratified intervals along the chromosome. This has the advantage of obviating the need to integrate and present maps with huge numbers of data-points on them.

6.3   Characters

One of the biggest benefits of having genetic maps is being able to localise genetic control of quantitative traits to specific regions of the genome, which requires genotyping and phenotyping of a random population of lines from a cross. For the same reasons that it is difficult for a plant breeder to select for important quantitative characters such as yield and quality, it is desirable to carry out phenotypic evaluation over a number of different environments. Thus, the phenotyping component is not trivial and ideally requires a population of inbred lines, either recombinant inbred lines (RILs) from a selfing series or doubled haploids (DH). Prior to the advent of molecular markers, manipulation of pairing in wheat genomes had been carried out to develop special chromosome stocks that carried substitutions of whole chromosomes or chromosome segments. Phenotypic analysis of these stocks enabled quantitative, and qualitative, variation to be firstly located to a chromosome and then to more specific regions. This elegant series of experiments, reviewed by Law et al.,[68] allowed much progress to be made in the genetic analysis of wheat. The development of the stocks was laborious and limited the range of germplasm that could be surveyed. The development of molecular maps meant that quantitative traits could be studied not only in a wider range of germplasm but also in more species.

6.3.1   QTL detection

Having assembled a set of genotypic and phenotypic data for a cross, the simplest way of detecting QTL is, for each marker in the genotype test, to classify the phenotypic data for each character into the parental marker groups. The means of each marker group can then be compared and tested for significance by analysis of variance or regression. Edwards et al.[69] detected QTL for a range of traits in maize using such an approach. The disadvantage of this approach is that it is laborious and only gives an association of marker loci with a character. In cases where neighbouring marker loci are far apart, the differences between the marker groups may not reach significance for a QTL of small effect and/or the QTL may be some distance from the marker.

Interval mapping[70] was developed to step along a genetic map at set intervals and, at each interval, test for the presence of a QTL. The test is based upon the phenotypic means of the marker classes and the distance between the markers. This approach gives a more precise location of QTLs and MAPMAKER/QTL software[71] has been developed to automate the procedure. This approach used maximum likelihood and could efficiently detect single QTL effects on a linkage group. In the presence of more than one QTL per linkage group, the method could either fail to detect any effect at all, if the loci from a parent were of opposite sign, or detect a 'ghost' QTL.[72] Haley and Knott[73] presented a least squares approach to interval mapping that gave similar results to maximum likelihood but was simpler to apply, could be adapted to a range of situations and was robust enough to detect multiple QTL in a linkage group.

QTLs of large effect could mask others of small effect and Jansen[74] proposed the use of co-factors to account for variation in other regions of the genome when scanning a target region. Jansen and Stam[75] proposed using the whole marker set as initial co-factors and then eliminating them in a backward elimination regression procedure but this can lead to over-parameterisation. Hackett[76] proposed a forward selection method to identify co-factors and this has been adopted by Utz and Melchinger.[77] Similarly Zeng[78] proposed the use of marker co-factors to account for the effect of one QTL when searching for another. The use of marker co-factors has been termed compound interval mapping and has been found, both by simulation and application to real data, to be more efficient at detecting QTL and also increases the precision of location. Software packages which implement this approach are MapQTL,[79] QTL Cartographer[80,81] and PLABQTL.[77] MQTL[82] uses background markers to account for QTLs in regions other than that currently being scanned in an approach termed 'simplified compound interval mapping'. PLABQTL and MQTL can also analyse data for a trait from more than one environment and determine which QTLs are consistent enough over environments to be termed main effect QTLs and which are effective in one or more environments (QTL × environment interactions). PLABQTL differs from MQTL in its approach to this problem as it looks for QTL in the overall means from different experiments and then attempts to fit this model to each environment in turn. This means that, as MQTL searches the accumulated data over all environments for QTL, it is more likely to detect QTL × environment interactions of the cross-over type[83] in addition to interactions of magnitude. Compound interval mapping is now the method of choice for QTL detection and choice of programme depends on objectives, availability and ease of use and interpretation.

Marker regression[84] has also been proposed as a method of QTL detection. This approach uses all markers in a linkage group to detect the presence of possible QTL and deviations of the observed values from predicted could indicate the presence of more than one QTL in a linkage group. This is a simple and easy to implement approach which has value in some situations but it does not detect QTL × environment interactions.

6.3.2   QTLs reported

A number of QTL studies in barley, oats and wheat have been published, largely centred around heading, height, yield and disease resistance. Some studies looking at malting quality parameters in barley and processing quality characters in wheat have also been published. Few comprehensive studies of agriculturally important traits have been published as most concentrate upon one or a few traits at a time. Barley appears to be an exception in that crosses have been studied for a range of agronomic, disease, yield and quality characters in North American,[48,85,86] European[16,58,87–90] and Australian[91] germplasm. No comparable studies have been published in either oats or wheat as most of the results in the public domain have concentrated on one or a few characters. For example, separate studies in wheat have focused on QTLs for plant height,[92] disease resistance,[93] flour viscosity[94] and frost tolerance.[95] There are fewer still published reports for oats but QTLs for resistance to Barley Yellow Dwarf Virus[62] and groat oil content have been detected.[96]

In a number of the QTL studies conducted in barley, a major gene with a marked agronomic effect has been segregating, either a dwarfing or a disease resistance gene. In many cases, the major genes have been associated with QTLs for other characters. For example, the *sdw1* dwarfing gene on chromosome 3H in spring barley has been associated with malting quality characteristics.[87,90,97] There are also a number of instances where QTL 'hot-spots' have been detected. For instance, the distal region of the short arm of chromosome 5H in Harrington × TR306 is associated with QTL for yield, maturity and some malting parameters amongst other characters.[48,86] The confidence intervals of QTLs are so large that it is still not possible to determine whether associations with major genes or 'hot-spots' are due to genuine pleiotropy or to tight linkage. The construction of special genetic stocks in which a series of isolines differ by small, neighbouring fragments of the genome[98] should help to resolve this question.

6.3.3   QTL validation

Before utilising QTL information in a breeding programme, some form of verification of effect is desirable. This problem has been addressed in several practical studies in barley. Hayes *et al.*[85] identified two QTL which, when combined, accounted for over 25% of the phenotypic variation in malt extract and some other malting quality parameters in the Steptoe × Morex mapping population. Han *et al.*[99] found that, in another sample of lines from Steptoe × Morex, the two QTL accounted for much less of the phenotypic variation for the characters. Romagosa *et al.*[100] came to the same conclusion in a study of yield in Steptoe × Morex. Beavis[101] has carried out extensive studies of QTL effects in maize and concluded that accurate estimation of QTL effect requires extremely large populations. Working with populations of 100–200 lines resulted in considerable bias in estimate of effect. Bias in estimation of QTL effect appears to be a genuine problem but need not detract from results already accumulated, provided that the methodology identifies the most important QTL.

6.4   Deployment of molecular markers

6.4.1   Varietal identification

Ainsworth and Sharp[102] proposed that RFLPs could be used in varietal identification but, as noted above, the overall level of polymorphism detected by RFLPs is relatively low and the technique is demanding. Terzi[103] demonstrated that RAPDs could also be used to distinguish between varieties of barley, oats and wheat but the disadvantage of this method is that the portions of the genome being sampled are unknown. Due to their high multiplex ratio, AFLPs are another possibility for cereal varietal discrimination.[104] The dominant nature and low polymorphic information content of AFLPs may, however, make them less useful in distinguishing between closely related cultivars and, like RAPDs, the genomic distribution of the markers is likely to be unknown. SSRs appear to be particularly useful for varietal discrimination, due to their multi-alleleic nature and high PIC values, and can provide unique genetic finger-prints of crop cultivars.[105] Russell et al.[106] used a panel of 11 SSRs to distinguish between 24 winter and spring barleys either on the UK National or Recommended Lists. In fact three different subsets of four SSRs were shown to be capable of differentiating between all 24 barleys. It should be noted that two of the winter barleys were derived by the same breeder from the same cross and were morphologically similar but differed at 4 of the 11 SSR loci surveyed.

6.4.2   Germplasm analysis

Cultivated germplasm

Molecular markers have a number of attractions for determining the genetic variability in a germplasm pool and, with the development of genetic maps, the level of polymorphism detected can be related to genomic regions. A multi-variate technique called 'principal co-ordinate analysis'[107] can be applied to data collected from a panel of genotypes to reveal the pattern of relationships between them. This relationship can be inspected graphically either by plotting the scores for the first two or three principal co-ordinates or, if they fail to account for a sufficiently large portion of the variation, a dendrogram. For barley, RFLPs have been used to differentiate between elite winter and spring European cultivars.[108,109] Within these groups, discrimination between either two- and six-rowed types is possible amongst the winter cultivars and between feed and malting cultivars amongst the spring cultivars.

RAPDs[110] and AFLPs[111,112] have also been used to examine the relationship between barley cultivars with similar results to those obtained from RFLPs. As for varietal discrimination, the disadvantages of using RAPDs and AFLPs is not knowing the chromosomal location of the markers. SSRs have a number of advantages in studying genetic relationships between genotypes as the multiple alleles that can be detected at any one locus give a better overall impression of the variation. Amongst European spring barley, Russell et al.[113] have shown that over 70% of all the alleles detected by a panel of 28 SSR loci distributed across all seven barley chromosomes were found in 17 'founder' genotypes which were largely old land-races, their early derivatives or disease resistance donors. Furthermore, more modern genotypes, as encompassed by a group released since 1985, possessed only 35% of the total alleles. This confirms the narrowing of the genetic base of modern barley, which Fischbeck[108] suggested was due to:

• the exploitation of relatively few land-races in plant breeding from the 1880s to 1920s
• a small number of cultivars that represented a significant advance and consequently featured heavily in subsequent breeding programmes
• the use of exotic germplasm largely as a source of disease resistance.

Whilst this dependence on 'founder' genotypes is fairly universal across the genome, Russell et al.[113] found that post-1985 spring barley cultivars did contain some novel alleles, notably in a region on chromosme 5H that may reflect selection for yield and/or quality.

Molecular analysis of wheat germplasm has not progressed as far as that of barley. The relatively low level of polymorphism detected by RFLPs in wheat[49,114] hindered progress and RAPDs were found to be unreliable due to primers amplifying different non-homologous sequences between genotypes.[115] Paull et al.[116] used a panel of mapped RFLPs to analyse a collection of Australian bread wheat genotypes to produce four major groupings that could be associated with the origin of the lines. They also found a number of differences between the observed and expected genotype based upon pedigree information. They concluded that, provided that the markers gave a broad, evenly-spaced genome coverage, the use of markers to estimate similarity was more meaningful than pedigree information as the latter did not account for selection and drift. Paull et al.[116] used the low level of polymorphism detected by RFLPs to argue that, where differences existed, they might well reflect the effects of phenotypic selection. In contrast, multivariate analysis of a sample of 11 winter wheat cultivars from Austria and Germany based on pedigree information, RFLPs, AFLPs and SSRs did not produce any meaningful groupings, nor were there any common trends between the four dendrograms.[117] Over all the pairwise combinations of the 11 genotypes, there were no significant correlations between the pairwise combinations of genetic similarities estimated by the three types of molecular markers and only the estimate produced by AFLP showed a significant correlation with pedigree.

Working with oats, O'Donoughue et al.[118] used principal co-ordinates analysis of RFLP data scored on a panel of 84 cultivars. They found two major groupings from the analysis, which generally corresponded to winter and spring cultivars, and that some sub-groups could be associated with breeding history. Whilst they found a significant correlation between genetic distance, as measured by RFLPs, and pedigree, like the studies of wheat and barley noted above, the correlation was small.

Exotic germplasm
ISSRs have been used not only to discriminate between winter and spring cultivated European barleys but also to highlight the distinctiveness of wild barley *Hordeum spontaneum*.[119] This has been explored further with the use of mapped SSRs, which has highlighted genetic bottlenecks that have occurred at specific regions of the barley genome, notably on chromosome 7H.[120] A survey of land-race material collected by ICARDA[121] shows that it is intermediate between *Hordeum spontaneum* and cultivated barley in terms of alleleic diversity. Greater diversity exists in the land-races than in *Hordeum spontaneum* in specific regions of the genome, much of it from novel alleles (J.R. Russell, pers. comm.).

In *Hordeum spontaneum*, Pakniyat et al.[122] analysed single accessions from 39 collection sites with AFLPs derived from 12 primer combinations. They found that the first three principal co-ordinates grouped the accessions according to the eco-geographic variables associated with the collection sites. The first principal co-ordinate differentiated between the sites on the basis of altitude whilst the second and third discriminated between longitude and latitude respectively. If one could then identify the phenotype that is favoured at a particular site, then such an analysis provides a powerful method of studying the direct action of natural selection.

6.4.3   Pedigree analysis

The multi-allelic nature of SSRs makes them extremely useful in analysing pedigrees of genotypes.[105] Swanston et al.[97] found a barley SSR, Bmac213, was linked within 7 cM of a major gene locus affecting non-production of epi-heterodendrin (a cyanogenic glycoside precursor important in some distilleries). They were able to trace the origin of this allele back through the pedigree of the spring barley cultivar Derkado to the cultivar Emir. Other derivatives of Emir were also non-producers of epi-heterodendrin and also possessed the same Bmac213 allele as Derkado. The origin of the Emir allele was probably the disease resistance donor Arabische, although the accession tested neither possessed the Derkado Bmac213 allele, nor was it a non-producer of epi-heterodendrin.[97] This was probably because the accession used in the breeding of Emir differed from that in the collection studied. This illustrates the general problem that material in collections, particularly less homogeneous material, may not accurately represent the variation in a particular genotype due to genetic drift and/or difficulties in ensuring that all the variation is passed on from generation to generation during multiplication and maintenance of accessions.

6.4.4   Marker-assisted selection

With the development of molecular maps, there are now many published associations of characters with markers that are close enough to be used in marker-assisted selection programmes (for a review of those published in wheat see Gupta et al.[123]). Theoretical studies have highlighted the benefits of applying marker-assisted selection in breeding programmes of the small-grained cereals[124] which suggest that marker-assisted selection should offer a substantial advantage over phenotypic selection for characters of low heritability under intense selection. The problem then becomes one of identifying robust QTLs that can be used in marker-assisted selection for such characters. Various selection strategies, including marker-assisted selection, have been compared for malt extract and some other malting quality characters[99] and yield[100] in barley. Both studies demonstrated that marker-assisted selection was effective, particularly when combined with phenotypic selection in the later stages of a breeding programme. A strategy of using genotypic selection to assemble a pool of 'elite' germplasm in which a number of desirable QTLs of major effect for traits that are difficult to measure followed by phenotypic selection to identify the best of these lines has a number of advantages:

- Selection is concentrated on lines that are most likely to meet the desired standard for yield and quality.
- The number of lines in the advanced stages of breeding programmes is reduced.
- Response from genotypic selection of QTL of small effect is, at best, likely to be negligible and therefore not cost-effective.

Marker-assisted selection in commercial breeding of wheat and barley is carried out, but generally for qualitative characters that are difficult to assess phenotypically. Markers can also be used to gain some knowledge about unknown germplasm, for example to choose parents for use in crossing. Alternatively, where progeny are known to be segregating for a qualitative character, marker-assisted selection may be applied at varying stages of the selection programme to identify those carrying the target gene. The Barley Yellow Mosaic Virus complex is one example where molecular markers have been deployed. The *rym4* resistance gene on barley chromosome 3HL was found to be flanked by the RFLPs MWG10 and MWG838. The latter was converted into a sequence tagged site (STS)[125] and Tuvesson *et al.*[126] demonstrated its use in molecular-assisted selection. The SSR Bmac29 is also linked to the *rym4* locus and is capable of differentiating not only between resistant and susceptible alleles but also between the *rym4* and *rym5* alleles.[127]

6.4.5   Backcross conversions

The benefits of using marker-assisted selction in backcrossing have been demonstrated theoretically.[128,129] Toojinda *et al.*[130] identified a source of barley stripe rust (*Puccinia striiformis*) resistance in exotic germplasm and then used a backcross conversion scheme to introgress the resistance into the locally adapted cultivar Steptoe. Progeny from the first backcross that carried the desired genotype were selected by using RFLP markers that were polymorphic and flanked the QTL. The selected BC1F1 genotypes will be heterozygous for the target segment and therefore doubled haploids were produced from them to produce inbred lines. The doubled haploids were then re-selected with the flanking RFLP markers to identify those that carried the QTL. AFLPs were used to screen the background genotype and showed that the percentage of the donor genome varied from 7 to 60%.[130] Under a normal backcrossing scheme, the percentage of the donor genome is expected to average 25% in a first backcross so some of the lines approached the equivalent of a third backcross. The capacity to identify such lines demonstrates the power of marker-assisted selection in a targeted backcrossing strategy. Selected backcross inbred lines can be produced within a two-year time-scale, regardless of whether the crop is winter- or spring-sown. The scheme is most suitable for one or two loci and, because it relies on selection of heterozygotes at the marker loci, works best with co-dominant markers for the target locus. If working with exotic germplasm, the relatively low levels of polymorphism detected by RFLPs should not be a problem. Given the increasing numbers of loci that cover the genomes of barley[22] and wheat[23] and the other advantages noted above, SSRs are the current marker of choice in targeted backcrossing schemes.

6.5   Future prospects

6.5.1   Candidate genes

Expressed sequence tags (ESTs) are developed by partially sequencing cDNA clones from target tissues. From the sequence information collected, searches can be made against publicly available databases and, given an adequate level of sequence homology, some idea of function can be derived. A big international effort is being made to assemble comprehensive EST databases for the *Triticeae* that will be publicly available.[131] At the same time, private companies are investing heavily in developing EST libraries for their 'core' crops but it is unlikely that much of their information will be made public. The challenge in both the private and public sector is to find ways of incorporating ESTs on genetic maps in order to determine whether any correspond to QTLs of interest, which would facilitate the change in emphasis in genetic maps from largely anonymous to known-function markers. Combining this information with QTL mapping will not only enable some functionality to be assigned to a particular QTL but also provide sequence data for its direct manipulation. Current QTL mapping methodology is, however, far too imprecise[132] to enable such assignations to be made unambiguously. Mapping large EST libraries also poses a number of problems, including choice of the most appropriate marker system. Some 5% of barley EST sequences have been found to possess SSRs (G.C. Machray, pers. comm.) but their informativeness has yet to be established. The identification of SNPs in ESTs appears to be more informative but identification currently requires a massive sequencing effort to reveal differences in panels of genotypes.

6.5.2   DNA chips

DNA sequences can now be assembled in micro-arrays over a small area – so-called DNA chips.[133] DNA chips are being advocated as suitable for a number of applications, including genotyping where they would be highly suitable for the detection of bi-allelic markers such as SNPs. The high level of automation that is possible with DNA chips means that it would be possible to carry out high through-put genotyping, which would enable marker-assisted selection on a large scale. DNA chip technology is still under development and its cost is not yet apparent. Whilst it appears promising, it will have to show a clear benefit over current approaches before it becomes a worthwhile investment purely for germplasm screening purposes. Apart from some obvious qualitative characters, the problem of identifying QTLs that are robust enough to warrant application of marker-assisted selection (noted above) will limit the amount of screening that could currently be done using DNA chips. If the technology was sufficiently cost-effective, then it may, however, be suitable for screening QTLs that are less robust.

6.5.3 Bioinformatics

The development of SSRs for a wide range of crop species has led to the possibility of not only examining genetical relationships through principal co-ordinates analysis but also using the map locations of the markers to produce a more meaningful graphical representation of the allelic constitution of genotypes.[134] Computer software has been developed to construct these graphical representations, e.g. Supergene,[135] but no package published so far can cope with the range of alleles that can be generated at one locus by SSRs. As SSRs are integrated into mapping studies of qualitative and quantitative characters, it becomes possible to assign values to specific SSR alleles. A variety of crossing strategies could then become possible by genotyping parental lines. Working within adapted germplasm, one could:

- assemble all the favourable QTL alleles for important characters that are difficult to screen for in cross combinations or
- make crosses between parents that both possess the major favourable QTL alleles and use conventional phenotypic selection to assemble the minor QTL alleles in a favourable combination.

Combining the information in such a database with genotypic information from the unadapted gene-pool would facilitate the identification of novel alleles in the region of important QTLs. These novel alleles may also represent novel variation for the character and can be introgressed through a marker-assisted backcrossing scheme to test whether or not they represent useful variation for deployment in plant breeding. Useful and novel phenotypic variation in wild relatives has already been detected in tomato[136] and rice,[137] although it remains to be thoroughly tested whether the QTLs from wild relatives represent new loci or variants at existing loci. The approach presented by Tanksley et al.[136] is interesting as it combines mapping with introgression in a methodology termed 'advanced backcrossing'. Its utility in the small-grained cereals discussed in this chapter remains to be evaluated.

6.5.4 Recombinant chromosome substitution lines (RCSLs)

One of the problems in establishing reliable QTL locations is that the estimate is usually based on very few recombinant lines and therefore it is not surprising that QTL effects vary in different studies, even when sampling from the same cross.[99,100] One solution to this problem is to create a series of recombinant chromosome substitution lines.[98] These are essentially a series of isogenic lines which differ only for a small (ideally contiguous) segment of chromosome from a donor line. These can then be replicated and widely tested along with the recipient cultivar to estimate the effect of individual segments in isolation. This approach can be applied in a sequential manner by an initial screen of lines with relatively large introgressed segments to identify candidates for further rounds of backcrossing to break up a segment into a series of smaller ones. This approach can therefore be used to derive more accurate estimates of QTL effect and position to refine the target segment to be used in marker-assisted selection. Another benefit of this approach is that it should help to distinguish between close linkage and pleiotropy, as well as providing material for gene cloning.

6.6   Conclusions

Molecular markers have undoubtedly revolutionised the genetic analysis of performance traits and can be applied in marker-assisted selection schemes. Markers have been successfully applied in plant breeding but there are few published examples. Those that have been published are largely concerned with the manipulation of qualitative characters that give a characteristic phenotype. The identification of QTL influencing important quantitative characters does not yet seem reliable enough to warrant the use of markers tagging QTLs in a marker-assisted selection programme. What is certain is that technological developments will continue and that much more information about the genetics, biochemistry and physiology of crop plants will emerge. The challenge is, and will continue to be, to integrate all the information with conventional phenotypic selection into efficient plant breeding programmes.

6.7   Sources of further information and advice

Much information is available through the World Wide Web. Graingenes (*http:// wheat.pw.usda.gov/*) is a database that contains a vast amount of information on mapping of markers and traits in barley, oats and wheat as well as images of the crops and their pathogens. Similar databases exist for maize (MaizeDB – *http:// www.agron.missouri.edu/*) and rice (Ricegenes – *http://ars-genome.cornell.edu/ rice/*). Demeters Genomes (*http://ars-genome.cornell.edu/index.html*) is another WWW source of a wide range of information. It has links not only to all the crop databases noted above but also to crop newsletters. UK CropNet (*http:// synteny.nott.ac.uk/*) contains databases on cereals and grasses assembled under a BBSRC initiative and also mirrors the above databases.

Useful books on plant breeding are the four in the Plant Breeding Series published by Chapman and Hall. The first is *Plant Breeding: Principles and Prospects* (eds M D Hayward, N O Bosemark and I Romagosa), the second is *Selection Methods in Plant Breeding* (eds I Bos and P Caligari), the third is *Statistical Methods for Plant Variety Evaluation* (eds R A Kempton and P N Fox) and the fourth is *Quantitative and Ecological Aspects of Plant Breeding* (eds J Hill, H C Becker and P M A Tigerstedt). Another suitable book is

Molecular biological tools in cereal breeding   127

Principles of Plant Breeding Second Edition by R W Allard and published by John Wiley and Sons Inc.

6.8   References

1. BINGHAM J and LUPTON F G H. Production of new varieties: an integrated approach to plant breeding. In Lupton F G H (ed.) *Wheat Breeding,* pp. 487–538. London, Chapman and Hall, 1987.
2. SILVEY V. The contribution of new varieties to cereal yield in England and Wales between 1947 and 1983. *Journal of the National Institute of Agricultural Botany*, 1986 **17** 155–68.
3. SNAPE J W and SIMPSON E. Uses of doubled haploid lines for genetical analysis in barley. In Asher M J C, Ellis R P, Hayter A M, Whitehouse R N H (eds) *Barley Genetics IV, Proceedings of the Fourth International Barley Genetics Symposium, Edinburgh, 1981,* pp. 704–9. Edinburgh, Edinburgh University Press, 1981.
4. POWELL W, THOMAS W T B, CALIGARI P D S and JINKS J L. The effects of major genes on quantitatively varying characters in barley. 1. The Gp*ert* locus. *Heredity*, 1985 **54** 343–8.
5. POWELL W, CALIGARI P D S, THOMAS W T B and JINKS J L. The effects of major genes on quantitatively varying characters in barley. 2. The *denso* and daylength response loci. *Heredity*, 1985 **54** 344–52.
6. THOMAS W T B, POWELL W and SWANSTON J S. The effects of major genes on quantitatively varying characters in barley. 4. The *GPert* and *denso* loci and quality characters. *Heredity*, 1991 **66** 381–99.
7. SWANSTON J S. The consequences for malting quality of *Hordeum laevigatum* as a source of mildew resistance in barley breeding. *Annals of Applied Biology*, 1987 **110** 351–6.
8. VON WETTSTEIN-KNOWLES P. Cloned and mapped genes: Current status. In Shewry P R (ed.) *Barley: Genetics, Biochemistry, Molecular Biology and Biotechnology*, pp. 73–98. Oxford, CAB International, 1991.
9. MCINTOSH R A, HART G E, DEVOS K M, GALE M D and ROGERS W J. Catalogue of gene symbols for wheat. In Slinkard A E (ed.) *Proceedings of the 9th International Wheat Genetics Symposium, Saskatoon, 1998 (5)* 235pp. Saskatoon, University Extension Press, University of Saskatchewan, 1998.
10. WORLAND A J, LAW C N, HOLLINS T W, KOEBNER R M D and GIURA A. Location of a gene for resistance to eyespot *Pseudocercosporella herpotrichoides* on chromosome 7D of bread wheat. *Plant Breeding*, 1988 **101** 43–51.
11. PAYNE P I. Genetics of wheat storage proteins and the effect of allelic variation on bread-making quality. *Annual Review of Plant Physiology,* 1987 **38** 141–53.
12. BOTSTEIN D, WHITE R L, SKOLNICK M and DAVIS R W. Construction of a genetic linkage map in man using restriction fragment length polymorph-

isms. *American Journal of Human Genetics*, 1980 **32** 314–31.

13. MOORE G, GALE M D, KURATA N and FLAVELL R B. Molecular analysis of small grain cereal genomes: current status and prospects. *Bio/Technology*, 1993 **11** 584–9.

14. WILLIAMS J G K, KUBELIK A R, LIVAK K J, RAFALSKI J A and TINGEY S V. DNA polymorphisms amplified by arbitrary primers are useful as genetic markers. *Nucleic Acids Research*, 1990 **18** 6531–6.

15. VOS P, HOGERS R, BLEEKER M, REIJANS M, VAN DE LEE T, HORNES M, FRIJTERS A, POT J, PELEMAN J, KUIPER M and ZABEAU M. AFLP: a new technique for DNA fingerprinting. *Nucleic Acids Research*, 1995 **23** 4407–14.

16. POWELL W, THOMAS W T B, BAIRD E, LAWRENCE P, BOOTH A, HARROWER B, McNICOL J W and WAUGH R. Analysis of quantitative traits in barley by the use of amplified fragment length polymorphisms. *Heredity*, 1997 **79** 48–59.

17. POWELL W, MORGANTE M, ANDRE C, HANAFEY M, VOGEL J, TINGEY S and RAFALSKI A. The utility of RFLP, RAPD, AFLP and SSR (microsatellite) markers for germplasm analysis. *Molecular Breeding*, 1996 **2** 225–38.

18. WAUGH R, BONAR N, BAIRD E, THOMAS B, GRANER A, HAYES P and POWELL W. Homology of AFLP products in three mapping populations of barley. *Molecular & General Genetics*, 1997 **255** 311–21.

19. WEBER J L and MAY P E. Abundant class of human DNA polymorphisms which can be typed using the polymerase chain-reaction. *American Journal of Human Genetics*, 1989 **44** 388–96.

20. LAGERCRANTZ U, ELLEGREN H and KAKANUGA T. The abundance of various polymorphic microsatellite motifs differs between plants and vertebrates. *Nucleic Acids Research*, 1993 **21** 1111–15.

21. WEBER J L. Informativeness of human (DC-DA)N.(DG-DT)N polymorphisms. *Genomics*, 1990 **7** 524–30.

22. RAMSAY L, MACAULAY M, DEGLI IVANISSIVICH S, MACLEAN K, CARDLE L, FULLER J, EDWARDS K, TUVESSON S, MORGANTE M, MASSARI A, MAESTI E, MARMIROLI N, SJAKSTE T, GANAL M, POWELL W and WAUGH R. A simple sequence repeat-based linkage map of barley. *Genetics,* 2000 (In Press).

23. ROEDER M S, KORZUN V, WENDEHAKE K, PLASCHKE J, TIXIER M-H, LEROY P and GANAL M W. A microsatellite map of wheat. *Genetics*, 1998 **149** 2007–23.

24. ZIETKIEWICZ E, RAFALSKI A and LABUDA D. Genome fingerprinting by simple sequence repeat (SSR)-anchored polymerase chain reaction amplication. *Genomics*, 1994 **20** 176–83.

25. PROVAN J, THOMAS W T B, FORSTER B P and POWELL W. *Copia*-SSR: A simple marker technique which can be used on total genomic DNA. *Genome*, 1999 **42** 363–6.

26. KALENDAR R, GROB T, REGINA M, SUONIEMI A and SCHULMAN A. IRAP and REMAP: two new retrotransposon-based DNA fingerprinting techniques. *Theoretical & Applied Genetics*, 1999 **98** 704–11.

27. WAUGH R, McLEAN K, FLAVELL A J, PEARCE S R, KUMAR A, THOMAS W T B and

POWELL W. Genetic distribution of Bare-1-like retrotransposable elements in the barley genome revealed by sequence-specific amplification polymorphisms (S-SAP). *Molecular and General Genetics*, 1997 **253** 687–94.

28. BLAIR M W, PANAUD O and McCOUCH S R. Inter-simple sequence repeat (ISSR) amplification for analysis of microsatellite motif frequency and fingerprinting in rice (*Oryza sativa* L.). *Theoretical & Applied Genetics*, 1999 **98** 780–92.

29. SMITH L. Genetics and cytology of barley. *Botanical Reviews*, 1951 **17** 1–51, 133–202, 285–355.

30. NILAN R A. The cytology and genetics of barley 1951–1962. *Monographic Supplement No 3 Research Studies*, 278pp. Washington, Washington State University Press, 1964.

31. JENSEN J and JORGENSEN J H. The barley chromosome 5 linkage map. I. Literature survey and map estimation procedure. *Hereditas*, 1975 **80** 5–16.

32. JENSEN H P, JENSEN J and JORGENSEN J H. On the genetics of the 'laevigatum' powdery mildew resistance in barley. In Munck L (ed.) *Barley Genetics VI (I), Proceedings of the Sixth International Barley Genetics Symposium, Helsingborg, 1991*, pp. 593–5. Copenhagen, Munksgaard, 1992.

33. HILBERS S, FISCHBECK G and JAHOOR A. Localisation of the laevigatum resistance gene MlLa against powdery mildew in the barley genome by the use of RFLP markers. *Plant Breeding*, 1992 **109** 335–8.

34. GIESE H, HOLM-JENSEN A G, JENSEN H P and JENSEN J. Localization of the laevigatum powdery mildew resistance gene to barley chromosome 2 by the use of RFLP markers. *Theoretical & Applied Genetics*, 1993 **85** 897–900.

35. AL-BANNA M K S, JINKS J L and POONI H S. The contribution of pleiotropy at the mop loci to continuous variation in *Nicotiana rustica*. *Heredity*, 1984 **52** 95–102.

36. LANDER E S, GREEN P, ABRAHAMSON J, BARLOW A, DALY M J, LINCOLN S E and NEWBURG L. MAPMAKER: an interactive computer package for constructing primary genetic linkage maps of experimental and natural populations. *Genomics*, 1987 **1** 174–81.

37. STAM P. Construction of integrated genetic linkage maps by means of a new computer package: JOINMAP. *Plant Journal*, 1993 **3** 739–44.

38. STAM P and VAN OOIJEN J W. Joinmap™ version 2.0: Software for the calculation of genetic linkage maps. Wageningen, CPRO-DLO, 1995.

39. LIU B H and KNAPP S J. Gmendel: A program for mendelian segregation and linkage analysis of individual or multiple progeny populations using log likelihood ratios. *Journal of Heredity*, 1990 **81** 407.

40. LINDE-LAURSEN I. Giemsa C-banding of barley chromosomes. In Gaul H (ed.) *Barley Genetics III. Proceedings of the Third International Barley Genetics Symposium, Garching 1975* p. 304. Munich, Verlag Karl Thiemig, 1976.

41. LINDE-LAURSEN I. Linkage map of the long arm of barley chromosome 3 using C-bands and marker genes. *Heredity*, 1982 **49** 27–35.

42. GALE M D and REES H. The production and assay of segmental substitution lines in barley. *Genetical Research, Cambridge*, 1971 **17** 245–56.

43. LAURIE D A, PRATCHETT N, DEVOS K M, LEITCH I J and GALE M D. The distribution of RFLP markers on chromosome 2(2H) of barley in relation to the physical and genetic location of 5S rDNA. *Theoretical & Applied Genetics*, 1993 **87** 177–83.

44. KUNZEL G and KORZUN L. Physical mapping of cereal chromosomes, with special emphasis on barley. In Scoles G and Rossnagel B (eds) *V International Oat Conference & VII International Barley Genetics Symposium Proceedings, Invited Papers*, pp. 197–206. Saskatoon, University Extension Press, University of Saskatchewan, 1996.

45. HEUN M, KENNEDY A E, ANDERSON J A, LAPITAN N L V, SORRELLS M E and TANKSLEY S. Construction of a restriction fragment length polymorphism map for barley (*Hordeum vulgare*). *Genome*, 1991 **34** 437–47.

46. GRANER A, JAHOOR A, SCHONDELMAIER J, SIEDLER H, PILLEN K, FISCHBECK G, WENZEL G and HERRMANN R. Construction of an RFLP map of barley. *Theoretical & Applied Genetics*, 1991 **83** 250–6.

47. KLEINHOFS A, KILIAN A, MAROOF M A S, BIYASHEV R M, HAYES P, CHEN F Q, LAPITAN N, FENWICK A, BLAKE T K, KANAZIN V, ANANIEV E, DAHLEEN L, KUDRNA D, BOLLINGER J, KNAPP S J, LIU B, SORRELLS M, HEUN M, FRANCHOWIAK J D, HOFFMAN D, SKADSEN R and STEFFENSON B J. A molecular, isozyme and morphological map of the barley (*Hordeum vulgare*) genome. *Theoretical & Applied Genetics*, 1993 **86** 705–12.

48. TINKER N A, MATHER D E, ROSSNAGEL B G, KASHA K J, KLEINHOFS A, HAYES P M, FALK D E, FERGUSON T, SHUGAR L P, LEGGE W G, IRVINE R B, CHOO T M, BRIGGS K G, ULLRICH S E, FRANCKOWIAK J D, BLAKE T K, GRAF R J, DOFING S M, MAROOF M A S, SCOLES G J, HOFFMAN D, DAHLEEN L S, KILIAN A, CHEN F, BIYASHEV R M, KUDRNA D A and STEFFENSON B. Regions of the genome that affect agronomic performance in two-row barley. *Crop Science*, 1996 **36** 1053–62.

49. CHAO S, SHARP P J, WORLAND A J, WARHAM E J, KOEBNER R M D and GALE M D. RFLP-based genetic maps of wheat homologous group 7 chromosomes. *Theoretical & Applied Genetics*, 1989 **78** 495–504.

50. GILL K S, LUBBERS E L, GILL B S, RAUPP W J and COX T S. A genetic linkage map of *Triticum tauschii* (DD) and its relationship to the D genome of bread wheat (AABBDD). *Genome*, 1991 **34** 362–74.

51. LIU Y-G and TSUNEWAKI K. Restriction fragment length polymorphism RFLP analysis in wheat. II. Linkage maps of the RFLP sites in common wheat. *Japanese Journal of Genetics*, 1991 **66** 617–34.

52. CADALEN T, BOEUF C, BERNARD S and BERNARD M. An intervarietal molecular marker map in *Triticum aestivum* L. Em. Thell. and comparison with a map from a wide cross. *Theoretical & Applied Genetics*, 1997 **94** 367–77.

53. O'DONOUGHUE L S, WANG Z, RODER M, KNEEN B, LEGGETT M, SORRELLS M E and TANKSLEY S D. An RFLP-based linkage map of oats based on a cross between two diploid taxa (*Avena atlantica* × *A. hirtula*). *Genome,* 1992 **35** 765–71.

54. O'DONOUGHUE L S, KIANIAN S F, RAYAPATI P J, PENNER G A, SORRELLS M E, TANKSLEY S D, PHILLIPS R L, RINES H W, LEE M, FEDAK G, MOLNAR S J, HOFFMAN D, SALAS C A, WU B, AUTRIQUE E and VAN DEYNZE A. A molecular linkage map of cultivated oat. *Genome,* 1995 **38** 368–80.

55. QI X, STAM P and LINDHOUT P. Comparison and integration of four barley genetic maps. *Genome,* 1995 **39** 379–94.

56. LANGRIDGE P, KARAKOUSIS A, COLLINS N, KRETSCHMER J and MANNING S. A consensus linkage map of barley. *Molecular Breeding,* 1995 **1** 389–95.

57. GIESE H, HOLM-JENSEN A G, MATHIASSEN H, KJAER B, RASMUSSEN S K, BAY H and JENSEN J. Distribution of RAPD markers on a linkage map of barley. *Hereditas,* 1994 **120** 267–73.

58. THOMAS W T B, POWELL W, WAUGH R, CHALMERS K J, BARUA U M, JACK P, LEA V, FORSTER B P, SWANSTON J S, ELLIS R P, HANSON P R and LANCE R C M. Detection of quantitative trait loci for agronomic, yield, grain and disease characters in spring barley (*Hordeum vulgare* L). *Theoretical & Applied Genetics,* 1995 **91** 1037–47.

59. BECKER J, VOS P, KUIPER M, SALAMINI F and HEUN M. Combined mapping of RFLP and AFLP markers in barley. *Molecular and General Genetics,* 1995 **249** 65–73.

60. QI X, STAM P and LINDHOUT P. Use of locus-specific AFLP markers to construct a high-density molecular map in barley. *Theoretical & Applied Genetics,* 1998 **96** 376–84.

61. PENNER G A, ZIRINO M, KRUGER S and TOWNLEY-SMITH F. Accelerated recurrent parent selection in wheat with microsatellite markers. In Slinkard A E (ed.) *Proceedings of the Ninth International Wheat Genetics Symposium, Saskatoon, 1998,* pp. 131–4. University Extension Press, Extension Division, University of Saskatchewan, Saskatoon, Canada, 1998.

62. JIN H, DOMIER L L, KOLB F L and BROWN CM. Identification of quantitative loci for tolerance to barley yellow dwarf virus in oat. *Phytopathology,* 1998 **88** 410–15.

63. BECKER J and HEUN M. Barley microsatellites: allele variation and mapping. *Plant Molecular Biology,* 1995 **27** 835–45.

64. THOMAS W T B, BAIRD E, FULLER J D, LAWRENCE P, YOUNG G R, RUSSELL J, RAMSAY L, WAUGH R and POWELL W. Identification of a QTL decreasing yield in barley linked to Mlo powdery mildew resistance. *Molecular Breeding,* 1998 **4** 381–93.

65. STEPHENSON P, BRYAN G, KIRBY J, COLLINS A, DEVOS K, BUSSO C and GALE M D. Fifty new microsatellite loci for the wheat genetic map. *Theoretical & Applied Genetics,* 1998 **97** 946–9.

66. KLEINHOFS A, KUDRNA D and MATTHEWS D. Co-ordinators report:

Integrating barley molecular and morphological/physiological marker maps. *Barley Genetics Newsletter*, 1998 **28** 89–91. http://wheat.pw.usda. gov/ggpages/bgn/28/ul1txt.html#9.

67. MATHER K. *The measurement of linkage in heredity*. London, Methuen, 1938.

68. LAW C N, SNAPE J W and WORLAND A J. Aneuploidy in wheat and its uses in genetic analysis. In Lupton F G H (ed.) *Wheat Breeding, pp. 71–107. London, Chapman and Hall, 1987.

69. EDWARDS M D, STUBER C W and WENDEL J F. Molecular-marker-facilitated investigations of quantitative-trait loci in maize. I. Numbers, genomic distribution and types of gene action. *Genetics*, 1987 **116** 113–25.

70. LANDER E S and BOTSTEIN D. Mapping Mendelian factors underlying quantitative traits using RFLP linkage maps. *Genetics,* 1989 **121** 185–99.

71. LINCOLN S, DALY M and LANDER E. *Mapping genes controlling quantitative traits with MAPMAKER/QTL Version 1.1: A tutorial and reference manual*. Cambridge MA, Whitehead Institute for Biomedical Research, 1993.

72. MARTINEZ O and CURNOW R N. Estimating the locations and the sizes of the effects of quantitative trait loci using flanking markers. *Theoretical & Applied Genetics,* 1992 **85** 480–8.

73. HALEY C S and KNOTT S A. A simple regression method for mapping quantitative trait loci in line crosses using flanking markers. *Heredity*, 1992 **69** 315–24.

74. JANSEN R C. Interval mapping of multiple quantitative trait loci. *Genetics,* 1993 **135** 205–11.

75. JANSEN R and STAM P. High resolution of quantitative traits into multiple loci via interval mapping. *Genetics*, 1994 **136** 1447–55.

76. HACKETT C A. Selection of markers linked to quantitative trait loci by regression techniques. In van Ooijen J W and Jansen J (eds) *Biometrics in Plant Breeding: Applications of Molecular Markers, Proceedings of the Ninth Meeting of the EUCARPIA Section Biometrics in Plant Breeding, 1994* pp. 99–106. Wageningen, CPRO-DLO, 1994.

77. UTZ H F and MELCHINGER A. PLABQTL: A program for composite interval mapping of QTL. *Journal of Agricultural Genomics*, 1996 **2** http:// www.ncgr.org/ag/jag/papers96/paper196/indexp196.html.

78. ZENG Z B. Precision mapping of quantitative trait loci. *Genetics,* 1994 **136** 1457–68.

79. VAN OOIJEN J W and MALIEPAARD C. *MapQTL^{TM} version 3.0: Software for the calculation of QTL positions in genetic maps*. Wageningen, CPRO-DLO, 1996.

80. BASTEN C J, WEIR B S and ZENG Z B. Zmap – a QTL cartographer. In *Proceedings of the Fifth World Congress on Genetics Applied to Livestock Production: Computing Strategies and Software* **22** pp. 65–6. Guelph, Organising Committee 5th World Congress on Genetics Applied to Livestock Production, 1994.

81. BASTEN C J, WEIR B S and ZENG Z B. QTL Cartographer, version 1.13. Raleigh, North Carolina State University, 1999.

82. TINER N A and MATHER D. MQTL: software for simplified composite interval mapping of QTL in multiple environments. *Journal of Agricultural Genomics*, 1995 **1** http://www.ncgr.org/ag/jag/papers95/paper295/indexp295.html.

83. BAKER R J. Tests for crossover genotype-environmental interactions. *Canadian Journal of Plant Science*, 1988 **68** 405–10.

84. KEARSEY M J and HYNE V. QTL analysis: a simple 'marker regression' approach. *Theoretical & Applied Genetics*, 1994 **89** 698–702.

85. HAYES P M, LIU B H, KNAPP S J, CHEN F, JONES B, BLAKE T, FRANCKOWIAK J, RASMUSSON D, SORRELLS M, ULLRICH S E, WESENBERG D and KLEINHOFS A. Quantitative trait locus effects and environmental interaction in a sample of North American barley germ plasm. *Theoretical & Applied Genetics*, 1993 **87** 392–401.

86. MATHER D E, TINKER N A, LABERGE D E, EDNEY M, JONES B L, ROSSNAGEL B G, LEGGE W G, BRIGGS K G, IRVINE R B, FALK D E and KASHA K. Regions of the genome that affect grain and malt quality in a North American two-row barley cross. *Crop Science*, 1997 **37** 544–54.

87. THOMAS W T B, POWELL W, SWANSTON J S, ELLIS R P, CHALMERS K J, BARUA U M, JACK P, LEA V, FORSTER B P, WAUGH R and SMITH D. Quantitative trait loci for germination and malting quality characters in a spring barley cross. *Crop Science*, 1996 **36** 265–73.

88. BEZANT J, LAURIE D, PRATCHETT N, CHOJECKI J and KEARSEY M. Marker regression mapping of QTL controlling flowering time and plant height in a spring barley (*Hordeum vulgare* L.) cross. *Heredity*, 1996 **77** 64–73.

89. BEZANT J, LAURIE D, PRATCHETT N, CHOJECK J and KEARSEY M. Mapping QTL controlling yield and yield components in a spring barley (*Hordeum vulgare* L.) cross using marker regression. *Molecular Breeding*, 1997 **3** 29–38.

90. BEZANT J H, LAURIE D A, PRATCHETT N, CHOJECKI J and KEARSEY M. Mapping of QTL controlling NIR predicted hot water extract and grain nitrogen content in a spring barley cross using marker-regression. *Plant Breeding*, 1997 **116** 141–5.

91. LANGRIDGE P, KARAKOUSIS A, KRETSCHMER J, MANNING S, CHALMERS K, BOYD R, DAO LI C, ISLAM R, LOGUE S, LANCE R and SARDI. RFLP and QTL analysis of barley mapping populations. http://greengenes.cit.cornell.edu/WaiteQTL/. 1996.

92. CADALEN T, SOURDILLE P, CHARMET G, TIXIER M H, GAY G, BOEUF C, BERNAUD, S, LEROY P and BERNARD M. Molecular markers linked to genes affecting plant height in wheat using a doubled-haploid population. *Theoretical & Applied Genetics*, 1998 **96** 933–40.

93. FARIS J D, LI W L, LIU D J, CHEN P D and GILL B S. Candidate gene analysis of quantitative disease resistance in wheat. *Theoretical & Applied Genetics*, 1999 **98** 219–25.

94. UDALL J A, SOUZA E, ANDERSON J, ORRELLS M E and ZEMETRA R S. Quantitative trait loci for flour viscosity in winter wheat. *Crop Science,* 1999 **39** 238–42.

95. SUTKA J, GALIBA G and SNAPE J W. Inheritance of frost resistance in wheat (*Triticum aestivum* L.). *Acta Agronomica Hungarica,* 1997 **45** 257–63.

96. KIANIAN S F, EGLI M A, PHILLIPS R L, RINES H W, SOMERS D A, GENGENBACH B G, WEBSTER F H, LIVINGSTON S M, GROH S, O'DONOUGHUE L S, SORRELLS M E, WESENBERG D M, STUTHMAN D D and FULCHER R G. Association of a major groat oil content QTL and an acetyl-CoA carboxylase gene in oat. *Theoretical & Applied Genetics,* 1999 **98** 884–94.

97. SWANSTON J S, THOMAS W T B, POWELL W, YOUNG G R, LAWRENCE P E, RAMSAY L and WAUGH R. Using molecular markers to determine barleys most suitable for malt whisky distilling. *Molecular Breeding,* 1999 **5** 103–9.

98. PATERSON A H, DEVERNA J W, LANINI B and TANKSLEY S. Fine mapping of quantitative trait loci using selected overlapping recombinant chromosomes in an interspecies cross of tomato. *Genetics,* 1990 **124** 735–42.

99. HAN F, ROMAGOSA I, ULLRICH S E, JONES B L, HAYES P M and WESENBERG D. Molecular marker-assisted selection for malting quality traits in barley. *Molecular Breeding,* 1997 **3** 427–37.

100. ROMAGOSA I, HAN F, ULLRICH S E, HAYES P M and WESENBERG D. Verification of yield QTL through realized molecular marker-assisted selection responses in a barley cross. *Molecular Breeding,* 1999 **5** 143–52.

101. BEAVIS W D. QTL analyses: Power, precision, and accuracy. In Paterson A H (ed.) *Molecular dissection of complex traits.* Boca Raton, CRC Press, 1998.

102. AINSWORTH C C and SHARP P J. The potential role of DNA probes in plant variety identification. *Plant Varieties and Seeds,* 1989 **2** 27–34.

103. TERZI V. RAPD markers for fingerprinting barley, oat and triticale. *Journal of Genetics & Breeding,* 1997 **51** 115–20.

104. RIDOUT C J and DONINI P. Use of AFLP in cereals research. *Trends in Plant Science,* 1999 **4** 76–9.

105. McCOUCH S R, CHEN X, PANAUD O, TEMNYKH S, XU Y, CHO Y G, HUANG N, ISHII T and BLAIR M. Microsatellite marker development, mapping applications in rice genetic and breeding. *Plant Molecular Biology,* 1997 **35** 89–99.

106. RUSSELL J, FULLER J, YOUNG G, TARAMINO G, THOMAS W, MACAULAY M, WAUGH R and POWELL W. Discriminating between barley genotypes using microsatellite markers. *Genome,* 1997 **40** 442–50.

107. GOWER J C. Some distance properties of latent root and vector methods used in multivariate analysis. *Biometria,* 1966 **53** 588–9.

108. FISCHBECK G. Barley cultivar development in Europe – success in the past and possible changes in the future. In Munck L (ed.) *Barley Genetics VI (II), Proceedings of the sixth International Barley Genetics Symposium, Helsingborg, Sweden, 1991* pp. 885–901. Copenhagen, Munksgaard, 1992.

109. MELCHINGER A E, GRANER A, SINGH M and MESSMER M M. Relationships

among European barley germplasm: I. Genetic diversity among winter and spring cultivars revealed by RFLPs. *Crop Science,* 1994 **34** 1191–9.

110. TINKER N A, FORTIN M G and MATHER D E. Random amplified polymorphic DNA and pedigree relationships in spring barley. *Theoretical & Applied Genetics*, 1993 **85** 976–84.

111. ELLIS R P, McNICOL J W, BAIRD E, BOOTH A, LAWRENCE P, THOMAS W and POWELL W. The use of AFLPs to examine genetic relatedness in barley. *Molecular Breeding,* 1997 **3** 359–69.

112. SCHUT J W, QI X and STAM P. Association between relationship measures based on AFLP markers, pedigree data and morphological traits in barley. *Theoretical & Applied Genetics*, 1997 **95** 1161–8.

113. RUSSELL J R, ELLIS R P, THOMAS W T B, WAUGH R, PROVAN J, BOOTH A, FULLER J, LAWRENCE P, YOUNG G and POWELL W. A retrospective analysis of spring barley germplasm development from 'foundation genotypes' to currently successful cultivars. *Molecular Breeding,* 2000 (In press).

114. LIU Y G, MORI N and TSUNEWAKI K. Restriction fragment length polymorphism RFLP analysis in wheat. I. Genomic DNA library construction and RFLP analysis in common wheat. *Japanese Journal of Genetics*, 1990 **65** 367–80.

115. DEVOS K M and GALE M D. The use of random amplified polymorphic DNA markers in wheat. *Theoretical & Applied Genetics,* 1992 **84** 567–72.

116. PAULL J G, CHALMERS K J, KARAKOUSIS A, KRETSCHMER J M, MANNING S and LANGRIDGE P. Genetic diversity in Australian wheat varieties and breeding material based on RFLP data. *Theoretical & Applied Genetics,* 1998 **96** 435–46.

117. BOHN M, UTZ H F and MELCHINGER A E. Genetic similarities among winter wheat cultivars determined on the basis of RFLPs, AFLPs, and SSRs and their use for predicting progeny variance. *Crop Science*, 1999 **39** (1) 228–37.

118. O'DONOUGHUE L S, SOUZA E, TANKSLEY S D and SORRELLS M E. Relationships among North American oat cultivars based on restriction fragment length polymorphisms. *Crop Science*, 1994 **34** 1251–8.

119. DAVILA J A, SANCHEZ DE LA HOZ M P, LOARCE Y and FERRER E. The use of random amplified microsatellite polymorphic DNA and coefficients of parentage to determine genetic relationships in barley. *Genome*, 1998 **41** 477–86.

120. ELLIS R P, RUSSELL J, RAMSAY L, WAUGH R and POWELL W. Barley domestication – *Hordeum spontaneum*, a source of new genes for crop improvement. In *Scottish Crop Research Institute Annual Report 1998/99*, pp. 97–100. Dundee, Scottish Crop Research Institute, 1999.

121. CECCARRELLI S, GRANDO S and VAN LEUR J A G. Genetic diversity in barley landraces from Syria and Jordan. *Euphytica*, 1987 **36** 389–405.

122. PAKNIYAT H, POWELL W, BAIRD E, HANDLEY L L, ROBINSON D, SCRIMGEOUR C M, NEVO E, HACKETT C A, CALIGARI P D S and FORSTER B P. AFLP variation in wild barley (*Hordeum spontaneum* C. Koch) with reference to

salt tolerance and associated ecogeography. *Genome,* 1997 **40** 332–41.

123. GUPTA P K, VARSHNEY R K, SHARMA P C and RAMESH B. Molecular markers and their application in wheat breeding. *Plant Breeding,* 1999 **118** 369–90.

124. KNAPP S J. Marker-assisted selection as a strategy for increasing the probability of selecting superior genotypes. *Crop Science,* 1998 **38** 1164–74.

125. BAUER E and GRANER A. Basic and applied aspects of the genetic analysis of the *ym4* virus resistance locus in barley. *Agronomie,* 1995 **15** 469–73.

126. TUVESSON S, VON POST L, OHLUND R, HAGBERG P, GRANER A, SVITASHEV S, SCHEHR M and ELOVSSON R. Molecular breeding for the BaMMV/BaYMV resistance gene ym4 in winter barley. *Plant Breeding,* 1998 **117** 19–22.

127. GRANER A, STRENG S, KELLERMANN A, SCHIEMANN A, BAUER E, WAUGH R, PELLIO B and ORDON F. Molecular mapping and genetic fine-structure of the rym5 locus encoding resistance to different strains of the barley yellow mosaic virus complex. *Theoretical & Applied Genetics,* 1999 **98** 285–90.

128. HOSPITAL F, CHEVALET C and MULSANT P. Using markers in gene introgression breeding programs. *Genetics,* 1992 **132** 1199–210.

129. HOSPITAL F and CHARCOSSET A. Marker-assisted introgression of quantitative trait loci. *Genetics,* 1997 **147** 1469–85.

130. TOOJINDA T, BAIRD E, BOOTH A, BROERS L, HAYES P, POWELL W, THOMAS W, VIVAR H and YOUNG G. Introgression of quantitative trait loci (QTLs) determining stripe rust resistance in barley: an example of marker-assisted line development. *Theoretical & Applied Genetics,* 1998 **96** 123–31.

131. LANGRIDGE P, ANDERSON O, GALE M, GUSTAFSON P, McGUIRE P and QUALSET C. International Triticeae EST Cooperative (ITEC). http://wheat.pw.usda. gov/genome/

132. KEARSEY M J and FARQUHAR A G L. QTL analysis in plants: where are we now? *Heredity,* 1998 **80** 137–42.

133. HOHEISEL J D. Oligomer-chip technology. *Trends in Biotechnology,* 1997 **15** 465–9.

134. TANKSLEY S D, YOUNG N D, PATERSON A H and BONIERBALE M. RFLP mapping in plant breeding: new tools for an old science. *Bio/Technology,* 1989 **7** 257–64.

135. BOUTIN S R, YOUNG N D, LORENZEN L L and SHOEMAKER R C. Marker-based pedigrees and graphical genotypes generated by supergene software. *Crop Science,* 1995 **35** 1703–7.

136. TANKSLEY S D, GRANDILLO S, FULTON T M, ZAMIR D, ESHED Y, PETIARD V, LOPEZ J and BECK-BUNN T. Advanced backcross QTL analysis in a cross between an elite processing line of tomato and its wild relative *L. pimpinellifolium. Theoretical & Applied Genetics,* 1996 **92** 213–24.

137. XIAO J, LI J, GRANDILLO S, AHN S N, YUAN L, TANKSLEY S D and McCOUCH S. Identification of trait-improving quantitative trait loci alleles from a wild rice relative, *Oryza rufipogon. Genetics,* 1998 **150** 899–909.

7

Risk assessment and legislative issues

W. Cooper, formerly NIAB, Cambridge and J. B. Sweet, NIAB, Cambridge

7.1 Introduction

7.1.1 Current status of GM crop development

Since the first field trials of transgenic crops were conducted in the USA and France in 1986, there has been a rapid growth in activity with field trials being carried out globally (Table 7.1) involving at least 56 different crop species (Table 7.2). In 1999 the acreage of GM crop plants grown for commercial purposes world-wide was expected to reach 73 million acres, with crops grown mainly in the USA and Canada.

While Europe has led the way in terms of GM crop development and evaluation, the commercial situation, in the UK in particular, is very different.

Table 7.1 Releases of genetically modified organisms per country 1998

Country	%	Country	%
USA	70.45	Sweden	0.37
Canada	11.83	New Zealand	0.34
France	4.72	Denmark	0.31
Belgium	2.02	Brazil	0.28
UK	1.84	South Africa	0.17
Italy	1.71	Finland	0.11
Holland	1.47	Portugal	0.06
Spain	1.20	Russia	0.06
Japan	1.17	Bulgaria	0.05
Germany	0.89	Austria	0.03
Australia	0.88	Switzerland	0.03

Table 7.2   Genetically modified plant species (OECD figures, 1998)

African violet (*Saintpaulia ionantha*)	Maize (*Zea mays*) (38%)
Alfalfa (*Medicago sativa*)	Marigold (*Tagetes sp.*)
American Chestnut (*Castonea dentata*)	Melon (*Cucumis melo*)
Apple (*Malus domestica*)	Mustard (*Brassica juncea*)
Asparagus (*Asparagus officianalus*)	Oat (*Avena sativa*)
Barley (*Hordeum vulgare*)	Oilseed rape (*Brassica napus*) (13%)
Beet (*Beta vulgaris*)	Onion (*Allium cepa*)
Belladonna (*Astropa belladonna*)	Orange (*Citrus sp.*)
Broccoli, cauliflower and cabbage	Papaya (*Carica papaya*)
(*Brassica oleracea*)	Pea (*Pisum sativum*)
Forage rape (*B. oleracea* var. *acephala*)	Peanut (*Arachis hypogaea*)
Kale rape (*B. oleracea* var. *biennis*)	*Pelargonium sp.*
Brown mustard (*Brassica nigra*)	Pepper (*Capsicum annuum*)
Carnation (*Dianthus carophyllatus*)	Pine (*Pinus sp.*)
Carrot (*Daucus carotta*)	Pineapple (*Ananas comosus*)
European Chestnut (*Castanea sativa*)	Poplar (*Populus sp.*)
Chicory (*Cichorium intybus*)	Potato (*Solanum tuberosum*) (12%)
Chrysanthemum (*Chrysanthemum*	Rice (*Oryza sativa*)
morifolium)	Rose (*Rosa hybrida*)
Cotton (*Gossypium hirsutum*) (7%)	Silver Birch (*Betula pendula*)
Cranberry, European	Spruce *Picea sp.*
(*Vaccinium oxycoccus*)	Spruce, Norway *Picea abies*
Creeping bentgrass (*Agrostis stolonifera*)	Sorghum (*Sorghum bicolor*)
Cucumber (*Cucumis sativus*)	Sugar beet (*Beta vulgaris*) (2%)
Cucurbita texana	Sugar cane (*Saccharum officinarum*)
Cucurbita pepo	Sunflower (*Helianthus annuum*)
Currant (*Rubus idaeus*)	Sweet potato (*Ipomoea batatas*)
Eggplant (*Solanum melonogea*)	Sweetgum (*Liguidambar sp.*)
Ethiopian mustard (*Brassica carinata*)	Tamarillo (*Cyphomandra betacea*)
Eucalyptus (*Eucalyptus camaldulensis*)	Thale cress (*Arabidposis thaliana*)
Flax (*Linum usitatissium*)	Tobacco (*Nicotiana benthamiana*)
Gladiolus sp.	Tobacco (*Nicotiana tabacum*) (5%)
Grape (*Vitis vinifera*)	Tomato (*Lycopersicon esculentum*) (10%)
Kentucky Bluegrass (*Poa patensis*)	Turnip rape (*Brassica rapa*)
Kiwi fruit	Walnut (*Juglans sp.*)
(*Actinidia deliciosa* var. *deliciosa*)	Watermelon (*Citrullus lanatus*)
Lettuce (*Lactua sativa*)	Wheat (*Triticum aestivum*)
Lisianthus (*Eustoma grandiflorum*)	White mustard (*Sinapsis alba*)
Lupin (*Lupinus angustifolius*)	

Those species comprising the majority of releases are indicated by the relevant percentage of releases within the OECD.

To date the UK has approved 135 applications for release, but for research purposes only. Whilst there are an increasing number undergoing experimental and performance trials, no consents for release for commercial purposes have yet been granted. Commercialisation of the first GM variety is under review in response to mounting public opposition and demands for a five-year freeze until further experimental analysis satisfies concerns about GM crop safety.

The state of GM crop development in the UK can be summarised as follows: oilseed rape and maize are nearest to commercialisation; modifications include varieties tolerant to the herbicides glufosinate ammonium (Challenge variety) and glyphosate (Roundup variety). In addition oilseed rape varieties modified for expression of improved oil quality such as those expressing a high lauric acid content are also close to the market place.

A wide range of GM crops are currently in experimental trial including spring wheat (disease resistance), sugar beet (herbicide tolerant and altered carbohydrate metabolism), potato (altered carbohydrate, virus resistance) and maize (herbicide tolerant and insect resistance). Genetic engineering has also enabled higher yielding hybrid systems to be produced by the development of GM male sterile plants, a number of which are currently being tested for yield and overall performance. Cultivars of spring and winter oilseed rape, sugar beet, fodder beet and forage maize are currently being assessed in the UK's statutory National List (NL) trials. Inclusion of a variety onto the NL and the EC common catalogue is an essential precursor to commercialisation (see Section 7.4).

Crop development in the future is likely to continue with the production of varieties with improved pest and disease resistance. These developments are also likely to include plants compatible with effective weed control and environmentally friendly farming methods, and crops with tolerance to salinity, drought or frost. There is also likely to be an increasing emphasis on the development of varieties that are bred for processing purposes such as the production of novel oils, starches and high-value pharmaceutical compounds, for example, vaccines. GM crops are likely to be an important future development in providing a substitute for fossil fuels. There is good potential for utilising genetic engineering for the synthesis of plant-based alternatives to fossil fuels using a range of widely grown oilseed crops like oilseed rape to produce economically viable quantities of oil. GM crops may also provide substitutes for other non-renewable resources from which we derive, for example, many industrial oils used in the manufacture of plastics, detergents, inks and lubricants. One example is the isolation of the genes encoding the enzymes responsible for the synthesis of petroselinic acid, a fatty acid with potential for use in making detergents and nylon polymers. Further emphasis will also be placed on the development of designer health crops, resulting in better tasting, nutritionally enhanced or 'healthy option' crops. One example is potatoes with altered starch metabolism which have reduced oil uptake during cooking and, therefore, offer a healthy alternative to the traditional potato chip.

7.1.2   Concerns surrounding GM crops

The concept, let alone commercial reality, of genetically modified (GM) crops continues to be the cause of considerable concern to the public (see Section 7.5). Amongst the most frequently quoted concerns are the fears of gene escape to wild relatives leading to what has been termed gene pollution, the contamination of organically grown crops and the breakdown of disease/pest resistance in GM varieties. Other concerns relate to the safety of ingestion of GM ingredients by humans, for example the potential for developing allergenicity in a crop which was otherwise allergen free. GM ingredients containing antibiotic resistance genes have also met with public resistance due to fears associated with the potential for antibiotic resistant strains of bacteria developing via gene transfer in the gut of animals or even humans. Biotechnology companies and research organisations are responding to public pressure by developing GM varieties that no longer contain antibiotic resistance markers.

There is no doubt that the recent wave of public concern surrounding the safety and ethics of GM crops has overshadowed the significant potential benefits that GM technology has to offer for those involved in all parts of the food chain from primary producers/growers to the household consumer, as well as the potential environmental benefits associated with decreased spray applications. However, it is true that many GM crops may have impacts, including some very positive, upon agriculture and the environment while in some cases there may also be implications concerning food quality and safety. Ultimately there may also be ethical concerns for some sectors of society. Where there are benefits to be gained at a known or unknown risk, the question of risk assessment and subsequent risk management arises. If the potential of GM technology is to be realised, the quality, safety, benefits and ethical integrity of this new technology must be evaluated against the risks. Where the benefits are found to outweigh the risks the potential of transgenic technology needs to be realised by management under strict regulatory procedures and effective stewardship post-market release.

7.2   Risk assessment and avoidance: general principles

7.2.1   Principles of risk assessment

As previously mentioned, GM crops have a number of potential benefits for growers, processors and eventually the consumer, but it is also recognised that there are likely to be environmental impacts and implications for food quality and safety. For example, exploitation of novel GM pest-, disease- and herbicide-resistant crops will require different (often reduced) pesticide and herbicide applications. These modified management systems will have an impact upon current agricultural systems and the agricultural environment. Such impacts are best analysed by risk assessments.

The basic concepts of risk assessment for genetically modified crops are similar to those applied to chemical pesticides where the risk is equal to the frequency and the hazard. For example no exposure (frequency) would equate to zero hazard. Risk assessments study both the severity and extent of the hazard or damage as well as the likelihood and frequency at which the damage will occur. Risk is defined as:

$$\text{Risk (impact)} = \text{Frequency (exposure)} \times \text{Hazard}$$

Clearly the ideal situation would be one of zero risk. Since in reality the likelihood of risk is always greater than zero, acceptable risk levels for GM crops must be defined, as with all new technology. What is defined as acceptable is based upon cultural values and may well differ globally. Indeed the current climate of controversy surrounding GM crops signifies strong cultural differences between European and North American consumers in what is defined as acceptable levels of risk for the utilisation of GM crops.

While there are differences in the regulatory procedures controlling the development and commercialisation of GM crops in North America and Europe (Section 7.4), both systems apply the same broad principles to assessing the safety of GM crop usage for food, animal feed and in terms of environmental impact. The first step involves thoroughly assessing the procedure for modifying the plant tissue. In the UK, for example, the Advisory Committee on Genetic Modification (ACGM) is the regulatory authority responsible for contained use evaluation; that is, the initial experimental work 'contained' within the laboratory or glasshouse. The risk evaluation procedure must be specific to each product. Broadly drawn conclusions, for example based on inter-species comparisons, are unacceptable. Most importantly the information requested in a risk assessment must be derived scientifically, with experiments designed to provide clear, interpretable, unequivocal and reproducible results. A recent addition to the risk-assessment procedure has occurred in the UK, in response to public pressure, where there is now a move towards assessing the societal and cultural impacts of this new technology alongside the environmental and human health risks.

Risk assessment can be divided into four steps (Nickson and McKee 1998):

1.   problem formulation
2.   risk analysis
3.   risk characterisation
4.   risk management.

Problem formulation requires that all available information concerning the plant, the trait and the experimental information is gathered in the context of the most likely hazards, such as toxicity/allergenicity. Once all the data are available, they can be analysed for characterisation of the likelihood and/or severity of the risk. In the final phase of the assessment procedure, the acceptability or otherwise of the identified risk must be determined and effective plans set out for its management. The risk assessment procedure is an iterative one and must continue throughout the use of the product, including post-market monitoring.

In the case of GM crops there are a number of variables/risk types to consider including impacts on the agricultural environment, closely related species, insects and animals and human health. To analyse the consequences of GM crop impact upon the agricultural environment requires a detailed understanding of the characteristics of the GM crop in question. This involves determining which wild relative, if any, it may hybridise with and studying the management systems involved in growing the GM crop itself. It also involves recognising any potential effects on other GM or non-GM crops which are likely to be grown in rotation with the variety being assessed. As an example, GM herbicide tolerant (HT) crops will be treated with different herbicides, with different activity spectra, at different crop development stages, leading to effects on the botanical diversity in the GM-HT crop which are the product of the interaction between the GM crop and the herbicide treatment.

The nature of any hazard is dependent upon the characteristics of both the crop that is modified and of the GM trait. Risk assessments require measurement and study of the hazard or impact of both. Numerous studies have concentrated on measuring frequency phenomena such as gene flow and inter-specific hybridisation without considering the impact of the transgene when it has dispersed or introgressed into other populations or species. In addition the impact of the release of the GM plant will depend on the type and location of the environment into which it is being released. To be truly effective, risk assessments may have to be carried out for a range of locations as they are not necessarily transferable from one site, area, region or country to another.

7.2.2   Impact of plant species

Plants vary in the degree to which they are dominant or are invasive in certain environments and in their ability to disperse genes to different populations and species. They will therefore have different environmental impacts when genetically modified. For any particular country or region, plants can be classified as potentially being high, medium or low impact.

Plants in the high-impact group are generally hardy, perennial, competitive, open-pollinating and prolific having a wide range of relatives with which they hybridise and an ability to colonise a range of natural and semi-natural habitats. Examples include perennial rye grasses (*Lolium perenne*) and certain indigenous and introduced trees and shrubs that form a significant proportion of forests and woodlands, e.g. *Populus* spp. Modifications of these plants, which affect their competitiveness, could have significant impacts upon the ecology of a range of environments.

Medium-impact plants are open-pollinating, hybridise with some wild relatives, are prolific and colonise a limited range of habitats. Examples of such plants include oilseed rape, oats, sugar beet and rice, all of which have closely related wild relatives with which they hybridise and an ability to colonise disturbed ground. These plants and their close relatives rarely form climax populations except in particular environments such as coastal areas or in disturbed ground. Low-impact plants are usually annual or biennial species, are largely self-pollinating with few hybridising relatives that are poorly adapted (or not native) to the area in which they are cultivated. In the UK, examples include maize and sunflower.

It is important to appreciate that the impact of a plant species will depend upon the environment into which it is being released. Maize and potato are considered low-impact plants in England. However in Central and South America, where their centres of genetic diversity occur, their impact would be considered very high.

7.2.3 Impact of transgenes

Transgene expression in GM plants will have different impacts in different environments. Since genes often operate uniquely it is not easy to classify transgenes as having high or low impact. In addition their impact is also dependent upon the nature of the receiving environment (agricultural impact).

High-impact transgenes generally encode genetic modifications that improve the fitness of the GM plants by increasing their reproduction, competitiveness, invasiveness and/or persistence and will therefore also have the greatest environmental impact. Thus transformations that significantly increase plant productivity by overcoming constraints such as broad-spectrum pest, disease and stress tolerance will have the highest impact. Many pest- and disease-resistance genes will have effects on non-target species either directly or indirectly by altering relationships between pests and beneficial organisms. It is important that these non-target effects are thoroughly understood before commercialisation progresses.

Low-impact transgenes are genes that do not noticeably enhance the fitness of the modified plant so that the modified plant's role and behaviour in a given ecosystem is not altered. Examples would include genes that modify seed composition, e.g. high lauric acid genes in oilseed rape and high starch content genes in potato. However, in preparing a comprehensive risk assessment it would be important to confirm that low-impact genes might not, unintentionally, confer an environmental advantage. As an example, in the case of high starch content genes in potato, it would be important to assess that the transgenes do not significantly increase potato seed tuber over-wintering survival rates through enhanced frost resistance. In the case of oilseed rape, it would be important to ensure, for example, that there is no increase in the dormancy characteristics of oilseed rape which may confer enhanced soil survival characteristics.

7.2.4 Mechanisms of transgene transmission

Gene flow is an important consideration in evaluating the risks associated with growing GM crops. Transgene dispersal could lead to contamination of neighbouring crops, a particular worry since the UK organic authority amended its rules to include a zero tolerance to the presence of GM material. Transgene flow from crops to closely related wild relatives is also of concern as an environmental risk. Gene flow between different species is, however, not a new concept and has in fact been occurring between natural plant species, leading to a range of hybrids in the UK flora including amongst others the Salix, Lolium and Rumex genera (Daniels and Sheail 1999).

In order for gene transfer from one species of plant to another closely related wild relative to occur a number of barriers, both physical and genetic, must be overcome. These include dispersal (either of pollen or seed), longevity of the pollen grain, sexual compatibility, competition with other pollen sources and events post-fertilisation. Most gene dispersal occurs as a result of pollen transported either on the wind or via vectors such as bees (Ramsay *et al.* 1999) or, less commonly, by seed dispersal. The distances over which pollen dispersal occurs varies depending upon the plant species, the prevailing weather conditions, in the case of wind-borne pollen, or the insect vector (Moyes and Dale 1999). As discussed by Moyes and Dale (1999), although most studies have concentrated on the range of pollen dispersal, the survivability over time of the pollen grain is actually the most important aspect of potential gene transfer and cross-contamination.

Assuming that pollination is successful and gene transfer has occurred, the barriers to successful introgression of a gene from the original donor species to the recipient will be dependent upon what the gene might offer the recipient. If, for example, the gene induces a lethal effect, the seed of the recipient plant will die and gene introgression into the recipient species will go no further. If, however, the transgene confers a selective advantage such as cold tolerance, drought or disease resistance or the ability to thrive in low-light conditions, seed from the recipient plant will thrive. This is especially true for native species, with the greatest opportunity for transgene movement occurring within the crop-weed complex (Whitton *et al.* 1997). However, in assessing the scale of transgene movement, it is important to consider whether those plants containing genes conferring an adaptive advantage in the agricultural environment might lose that selective advantage in the differing environmental conditions outside of the farm field. If the transgene provides no selective advantage to the recipient plant, such as herbicide-resistance genes present in plants growing in an environment where herbicide spraying will not occur, the transgene will have a neutral impact upon the recipient species. There will be no increase in fitness of the population.

7.2.5   Multiple transgenes and transgene stability

One of the major issues surrounding GM crops containing multiple transgenes encoding a variety of traits is the question of stability of gene expression. Might the introduction of a second transgene affect expression of the original transgene and thus the phenotype of the GM variety? In particular, genetic homology between the two transgenes may cause down regulation of gene expression and suppression of the phenotype. How this effect is caused is complex and thought to be affected by factors such as the position of the transgene within the genome, i.e. point of insertion during the transformation procedure, transgene copy number within the genome and by other factors such as reproduction and even environmental conditions. The results may be unpredictable resulting in instability or silencing of gene expression (Senior and Dale 1996). The production of GM varieties involves evaluation of transgenic lines over a number of generations, during which any unstable lines would generally be identified and discarded. One possible exception to this would be instability arising from environmental interaction. This instability is also observed in conventionally bred varieties, providing a basis for further analysis of GM varieties (Qian *et al*. 1986).

From the perspective of risk assessment and environmental impact the most significant issue arises from gene flow between closely related species. Instability of gene expression generally leads to suppression of gene expression, in which case the phenotype of the GM variety would revert to the wild type, with no expression of the transgene. The implications for agronomic practice are significant, as suppression of gene expression would render a herbicide-tolerant GM variety susceptible to that particular herbicide, with consequent loss of yield if the farmer were to spray unwittingly. While the effect of transgene instability on the natural environment is likely to be minimal, there may be important issues at stake in the case of transgenic plants engineered to remove the synthesis of harmful toxins. In this situation suppression of gene expression arising from gene flow leading to multiple transgene insertions could prove a serious human or animal health problem if undetected.

7.3   Assessing the impact of genetically modified crops

7.3.1   Impact on agricultural systems

Genetic modification can have a range of impacts on agricultural systems and therefore will require specific agronomic management. The use of GM varieties would affect the nature of crop volunteers in subsequent crops and require alterations in volunteer management practices. The GM trait may also have an impact if it disperses to other crops and weeds through cross-pollination and seed dispersal. Low-impact genes such as herbicide tolerance, which have little impact on natural environments, become highly significant because of the changes in the herbicide usage required for their management. These herbicides will differ in the effect they have on plant and other species diversity in cropped fields.

Deployment of high-impact genes such as those encoding pest and disease resistance will result in reductions and changes in pesticide usage and thus offer opportunities to enhance diversity in cropped fields, especially if the transgene products are very specific to selected pests. However, it is important that the selection pressures they impose on pests and diseases do not encourage the development of virulent races of pests and pathogens and appropriate management systems are required in order to maintain durable resistance in the GM varieties.

7.3.2   Impact on uncultivated flora

Genetically modified crops may also have impacts on uncultivated and 'natural' environments. These environments may be affected by characteristics of crop and wild species induced by novel genetic constructs and their products. Risk assessments must therefore concentrate on whether the genetically modified characteristics of a GM crop and of similarly modified hybridising wild relatives are likely to change the behaviour of the plants or dependent flora and fauna in their environment, to the extent that ecological balances are altered.

7.3.3   Impact on insects and animals

The first successful example of using a foreign plant gene to confer resistance to insects was reported in 1987 (Hilder *et al.* 1987) and involved transformation of tobacco (*Nicotianum tabacum*) with the cow pea trypsin inhibitor (CpTi) gene. Since then there have been many reports of success in insect management using transgenic crop varieties.

The bacterial endotoxins isolated from *Bacillus thuringiensis (Bt)*, comprise one of several groups of proteins which have been shown to have insecticidal properties to a range of economically important insects. Transgenic crop varieties engineered with *Bt* resistance are already in commercial use in the USA and China, while a number of plant proteins, such as inhibitors of proteases, lectins and other digestive enzymes, are being evaluated for their efficacy as insect-resistance mechanisms (Gatehouse *et al.* 1998).

It is important that genes selected for the control of insect pests have acceptably little effect on non-target insects including predators of the target pest insects, in order to maintain insect diversity in GM crops. Clearly, if there is an effect upon predators that is comparable with current control practices then little benefit will accrue from the deployment of GM crops, a point made strongly by the Royal Society for the Protection of Birds in their submissions on GMOs (1997). Impact assessments are therefore required to examine the effects on non-target organisms in the crop environment.

Studies to evaluate the effect of transgenic plants expressing insect resistance on non-target species have provided, at best, equivocal and often controversial results which have served only to fuel the GM debate rather than provide hard scientific facts on which to base a thorough impact assessment.

Research into the impact of potato plants expressing the snowdrop lectin GNA upon 2-spot ladybirds which feed on the aphid *Mysus persicae* demonstrated that the ladybirds were affected adversely in terms of fecundity, egg viability and longevity (Birch *et al.* 1998). However, the authors point out that the effects may either be a direct result of ladybirds preying on aphids which have digested transgenic plant material containing the lectin, or may also be due to poor nutritional quality of the aphids themselves as a food source. Other studies involving the parasitic wasp *Eulophus pennicornis* and the tomato moth *Lacanobiua oleracea* demonstrated that the parasitic wasp was not affected when it parasitised moth larvae reared on transgenic potato plants expressing the snowdrop lectin GNA (Gatehouse *et al.* 1997).

More recently Losey *et al.* (1999) published a report indicating that pollen from transgenic *Bt*-resistant maize plants had a detrimental effect on the larvae of the non-target Monarch butterfly (*Danaus plexippus*), which is considered to be a sensitive indicator of environmental disturbance in the USA. Larvae, which normally feed on the leaves of the milkweed (*Asclepias curassavica*) plant were fed on leaves that had been dusted with unquantified amounts of pollen from the transgenic *Bt* maize plants. Results indicated that larval survival rate was only 56% compared to 100% survival for larvae fed on leaves dusted with untransformed pollen. Superficially these results indicate an unacceptable environmental impact from *Bt* maize. However, closer analyses have revealed a number of serious criticisms of the research, including the use of laboratory studies only, no-choice feeding regimes, lack of stringency, lack of quantification and the use of inappropriate controls (Hodgson 1999). The experiments were not conducted in the field so no *in vivo* data were available to confirm that (a) milkweeds occur in maize fields, and (b) that Monarch butterflies occur on these milkweeds bearing in mind the insecticide programme received by conventional maize. This once again reiterates the requirement for comprehensive risk assessments based on thorough science.

Schuler *et al.* (1999) have conducted research concerning the environmental effects of *Bt*-resistant GM oilseed rape on a non-target insect. The results demonstrated that the behaviour of non-target insects can also play a part in determining how *Bt* plants will affect their populations and should be considered when trying to evaluate the environmental impact of GM crops. Their laboratory-based experiments evaluated the ecological impact of the GM crop on the diamondback moth (*Plutella xylostella*), a pest that damages the oilseed rape crop, as well as the natural bio-control agent of the diamondback moth, a parasitic wasp (*Cotesia plutellae*), which kills the moths' caterpillars by laying its eggs in them. Results demonstrated that parasitoid wasp larvae that were oviposited in *Bt*-susceptible moth larvae not surprisingly died with their hosts. In contrast wasp larvae that had been oviposited in *Bt*-resistant moth larvae feeding on transgenic plants survived and demonstrated no adverse effects of exposure to the *Bt* toxins either as adults or in the development of their own larvae.

The research group then examined the behaviour of the female parasitic wasps in the presence of GM and non-GM leaves. It is known that the female wasps locate the host diamondback moth larvae using herbivore-induced volatiles released from the damaged plants. A wind-tunnel was used to compare the flight response of the wasp towards *Bt*-susceptible and *Bt*-resistant diamondback larvae which were allowed to feed on *Bt* leaves. The flight and feeding behaviour of each wasp was then measured. In this test, 79% of the parasitoids flew to the *Bt* leaves damaged by resistant moth larvae, with only 21% choosing *Bt* leaves damaged by susceptible larvae. The apparent lack of effect on the survival or host-seeking ability of the parasitic wasp suggested that *Bt* plants may have an environmental advantage over broad-spectrum insecticides.

7.3.4   Impact on human health

The potential for transferring genes from one unrelated species to another has caused concern that allergenicity may be introduced into a food source that was previously non-allergenic. An obvious example is that of the recent research programme by Pioneer Seeds where soya bean transformed with a gene from the Brazil nut was found to have allergenic properties. The research programme was halted before any field trials took place but public concern was heightened by reports of this work. All GM foods are now routinely tested for allergenicity using serological tests involving immuno-globulin antigens for specific allergenic proteins.

Transgene instability may be an important issue in the case of transgenic plants engineered to remove the synthesis of harmful toxins. In this situation suppression of gene expression arising from gene flow leading to multiple transgene insertions could prove a serious human or animal health problem if undetected (see Section 7.2.5).

The inappropriate choice of transgenes for achieving a desired trait may also have a serious impact on human health without adequate risk assessment. Lectins are a group of proteins known to have insecticidal properties that make them attractive candidates for the development of transgenic plants with resistance to the Homoptera insects. They are thought to work by binding carbohydrate side chains present in the gut wall resulting in inhibition of food absorption. As there is the potential of toxicity to humans, it is essential that extensive risk evaluation is required to establish any potential threats of toxicity. The need for such risk assessment is reflected in work on GM tomatoes (Noteborn et al. 1995) and was recently highlighted by reports from the Rowett Institute in Scotland, which indicated adverse immunological and nutritional effects from enhanced lecithin in GM potatoes (Ewen and Pusztai 1999, Fenton et al. 1999). However, aspects of the former of these reports in particular were strongly criticised by a number of reputable scientific bodies as being unsubstantiated and have highlighted the need for agreed methodology in this field of research so that conclusive results can be acquired (Kuiper et al. 1999).

There has also been concern about genetically modified ingredients containing antibiotic resistance genes used to select transformed cells prior to the regeneration of transgenic plants. Use of these genes raises the potential for antibiotic-resistant strains of bacteria to develop via horizontal gene transfer in the gastro-intestinal tract of animals or even humans (Harding and Harris 1997). This possibility is not thought to be a major hazard since the antibiotic-resistance genes most often used for plant transformation themselves come from bacteria. They encode resistance against antibiotics rarely used in medicine such as kanamycin, against which a large percentage of gut microflora is already resistant.

However, given the risk, however small, of producing more antibiotic-resistant bacteria, techniques are being developed that will enable selectable markers to be removed from crop plants after the transformation process. Alternative selectable markers, not based on antibiotic selection, are also being tested, for example a mannose permease that allows the use of mannose, a sugar not normally available to plant metabolism, as a carbon source during plant regeneration.

Risk assessment methodology will also have to be adjusted for food plants which are modified to improve nutritional and other qualities, a major area for current and future research. Target traits include, for example, improving the nutritional value of proteins, increasing the concentrations of oils low in saturated fats, or fortification with micronutrients or antioxidants. Food plants modified in this way must undergo extensive toxicological and nutritional assessment with a combination of *in-vitro* and *in-vivo* tests, as currently required for all novel foods by the EU, for example. In the case of genetic modification, however, particular attention needs to be given to the detection and characterisation of potential unintended effects of modification. Inferences about such effects can no longer be based solely on chemical analysis of single macronutrients and micronutrients, and known crop-specific antinutrients or toxins. New methods have been developed to screen for potential alterations in the metabolism of the modified organism by such methods as:

- analysis of gene expression (monitored, for example, by microarray technology or mRNA fingerprinting)
- overall protein analysis (proteomics)
- secondary metabolite profiling.

Studies using these designs will need to be designed carefully to take account of the complexity of foods (OECD 1996, Noteborn *et al.* 2000, Van Hal *et al.* 2000).

7.4   How is biotechnology regulated?

7.4.1   International picture

At an international level, a joint FAO/WHO initiative has agreed a basic framework for evaluating the safety of transgenic food crops (FAO/WHO 1996). Key aspects of any evaluation should include:

- characterisation of the new gene product
- identification of alterations in concentrations of nutrients and known toxicants
- assessment of the potential allergenicity of the gene product
- assessment of the impact of gene transfer between plants and the gut microflora of animals and humans.

An early basis for evaluation has been the concept of 'substantial equivalence'; how a new food compares with a conventional product that experience and use has proved to be safe for consumption (OECD 1993).

Although there is some consensus on approach, there are as yet no internationally agreed procedures or advisory bodies in place to regulate the release, either experimentally or for commercialisation, of GM crops. The approach to the regulation concerning releases to the environment differs between countries, although at the time of writing discussions concerning a global regulatory system are under way. The one exception is the European Union which has co-operated to form an agreed framework of legislation through EC Directive 90/220 *on the deliberate release into the environment of genetically modified organisms*. This directive, while ultimately seeking a unanimous decision concerning consent to release for commercialisation, leaves the individual member state free to decide whether to grant permission for release for experimental purposes.

7.4.2   UK regulatory process

For a new GM variety to be released into the environment in the UK, an application for consent to release, which includes a risk assessment, must be made. Interestingly, the UK Competent Authority, the Department of the Environment, Transport and the Regions (DETR) has decided to adopt a policy of transparency in these evaluations. As such, all information on every application for consent to release (except commercially sensitive data), is placed on the Public Register in summary form and is available in more detail on application to the DETR. In addition any application for release for experimental or commercial purposes, without exception, must be advertised publicly in the vicinity of the release site (normally in a local newspaper) and the consent approval, conditions of release and location of the release are made available via the Public Register.

Process in detail

In the UK responsibility for the safe release and subsequent monitoring of GM crops lies with the Department of the Environment, Transport and the Regions (DETR), which is recognised within the EU as the UK's Competent Authority. The DETR as the lead department, in consultation with all the other relevant UK departments, has outlined very specifically the requirements for anyone wishing to grow and market GM varieties in the UK. The applicant must notify the Health and Safety Executive (HSE) of their intention to initiate research work at the laboratory stage and produce an assessment of any risks to the environment or human health. The HSE will then consult the Advisory Committee on Genetic Modification (ACGM). The ACGM is responsible for regulating the contained use (laboratory or glasshouse) of Genetically Modified Organisms under the Genetically Modified Organisms (Contained Use) Regulations 1996 which came into force on 27 April 1996. Along with European Directive 90/219 the regulations establish a notification scheme for activities with genetically modified organisms (GMOs) according to their potential harmfulness to people or to the environment.

If the variety is to be grown outside for field trials a detailed application and environmental risk assessment must be completed by the applicant and submitted to the DETR. Applications for consent to release a GM variety into the environment are evaluated depending upon the nature of the release application, i.e. whether the request is for release for experimental and evaluation purposes (Part B release) or the applicant is requesting permission to release the GM variety for commercialisation (Part C release).

The risk assessment needs to contain information concerning the nature of the inserted transgene(s) and a detailed characterisation of the genetically modified plant arising from transformation with the transgene(s). The following information is sought:

- source of the transgene(s) and additional sequences
- details of the modification
- method of transformation
- level of interaction
- level of transgene(s) expression
- environmental interaction
- impact on human health
- management considerations for the release site.

The DETR is advised by the Advisory Committee on Releases to the Environment (ACRE), a statutory committee responsible for considering all Part B releases (release for experimental purposes only). A second committee, the Advisory Committee on Novel Food Processes (ACNFP), responsible to MAFF, advises on food safety issues. There is yet another committee, also answerable to MAFF, which is responsible for considering issues surrounding the use of GM varieties in animal feed. In evaluating GM issues all three committees may be required to provide advice and guidance.

ACRE's judgement on whether or not to grant consent for release for experimental purposes is based on evaluating the impact of answers to the following: is it possible that the inserted gene(s) may

- make the modified crop more persistent
- make the modified crop more invasive
- make the modified crop more undesirable to living organisms or the environment
- be transferred to other organisms?

If an application is successful a consent from the Secretary of State for the Environment will be given allowing the varieties to be grown outside for experimental purposes (Part B consent).

If the variety is progressed from experimental release to commercial field trials then the ACNFP would consider the new variety for food safety purposes under the EC Novel Foods regulation. The ACNFP holds responsibility for evaluating the safety of any novel foods submitted for approval under the EC Novel Foods and Novel Food Ingredients Regulation (258/97). This regulation, which came into effect on 15 May 1997, defines a novel food as food that has not been used for human consumption to a significant degree within the

Community. What this means in reality is that foods which have been sold in one or more member states before 15 May 1997 will not generally be covered by the regulation. This includes food ingredients obtained from the herbicide tolerant GM soya beans and insect-resistant maize, which were approved for sale under the Deliberate Release Directive 90/220/EEC. These food ingredients, which are on sale in a variety of processed foods, are not regarded as novel foods but are covered under EC regulation 1139/98 which requires the labelling of food ingredients containing GM material derived from soya or maize.

The ACNFP may then seek advice from the Food Advisory Committee (FAC), Committee on Toxicity of Chemicals (COT) and the Committee on Medical Aspects of Food Policy (COMA). A marketing consent is also required before the crops can be grown commercially, which requires another detailed submission to the DETR for consideration by ACRE. Once these committees approve an application their assessment is passed to other EU member states for consideration. Finally, if the new GM variety has been awarded Part C consent by the Department of the Environment, Transport and the Regions (DETR) the new variety may go forward for statutory evaluation for plant variety registration (see below) in the member state with a view to commercialisation. If no objections are raised and the new variety performs well, the product could be marketed, provided it is labelled in compliance with the new EU food labelling laws.

In the UK any new crop variety, including genetically modified varieties, targeted for commercialisation must comply with the regulatory requirements under the UK seeds legislation, the UK Environmental Protection Act 1990 and the GMO (Deliberate Release) Regulations 1992 and 1995, in addition to the previously mentioned EC Directive 90/220.

GM and non-GM varieties alike must also undergo statutory evaluation for registration on the National List, leading to plant breeder's rights and eventually the European Common Catalogue. This requires evaluation of the Distinctness, Uniformity and Stability (DUS) of a variety and its Value for Cultivation and Use (VCU). A GM variety will not be added to the UK National List until it receives full Part C consent from the EC, covering all issues relating to the environment, human health, animal feed and human food. In other words, full Part C consent is the prerequisite for variety registration and commercialisation. Once the new GM variety has successfully passed through all these evaluation phases a breeder may consider commercial development of the new variety.

The current regulatory system in the EU based on Directive 90/220/EEC has been heavily criticised due to the length of time taken for applications for consent to release to be evaluated. The delays arise as a result of the evaluation process, which looks at each application, based on the process (how the GM crop is derived) rather than the end product (modified trait). The product-orientated approach operated in the USA and Canada results in a much faster regulatory procedure. Discussions are currently under way aimed at reviewing the EU procedures.

7.4.3   Regulation in the USA and Canada

The USA and Canada have adopted a different approach to the regulatory process from that taken by the EC where the regulations relate to the process of modification. In the USA the regulation relates to the product and not the process, while in Canada the regulation relates to any plant with novel traits, which may include plants developed by approaches other than genetic modification.

The product-driven approach in North America means that once a product is proven safe it is no longer regulated. Similar releases by other organisations employing the same crop species/trait modification are subject to notification only, resulting in a fast track for proven modifications. The criteria used to determine safety to the environment and human health are equally rigorous to those considered in Europe, but the product-driven approach leads to faster approval and commercialisation.

The agencies responsible for regulating biotechnology in the USA are the US Department of Agriculture (USDA), the Environmental Protection Agency (EPA) and the Food and Drug Administration (FDA). The USDA has responsibility for regulations concerning plant pests and plants. Within the USDA the Animal and Plant Health Inspection Service (APHIS) is responsible for protecting US agriculture from pests and diseases. Notification is required for the introduction of a genetically modified organism that is considered to be a plant pest, such as a modified plant with weedy characters or a modified pathogen.

The EPA is responsible for regulating novel micro-organisms, microbial and plant pesticides and new uses of existing pesticides including use on GM herbicide tolerant crops. The FDA has regulatory responsibility for food, feed, and food additives derived from new plant varieties. The FDA policy requires that genetically modified ingredients meet the same safety standards as required for all other foods. Products are regulated according to their intended use, which means that some products will fall under more than one regulatory agency. For example, viral resistance in a food crop would be evaluated for safety to grow by the USDA, environmental safety by the EPA and by the FDA for safety to human health.

7.4.4   Food labelling

Novel foods such as GM foodstuffs are, where feasible, assessed in comparison with the foods they will replace. The process of substantial equivalence, as this assessment is known, was developed by the World Health Organisation (WHO) and the Organisation for Economic Co-operation and Development (OECD) and is an internationally accepted evaluation procedure. Evaluation of a GM foodstuff involves a wide range of information including agronomic aspects of the plant material and detailed information concerning the nutritional composition of the material. The evaluation looks at the intentional effects of the modification as well as any unintentional effects.

In addition to standard labelling requirements for food, the EC Novel Foods and Novel Food Ingredients Regulation outlines specific labelling requirements intending the consumer to be notified of:

- Any characteristic or food property which renders a novel food no longer equivalent (practically interpreted as meaning where GM material, either DNA or protein, is present in the food or ingredient) to an existing food or food ingredient.
- The presence in the novel food or food ingredient of material which is not present in an existing equivalent foodstuff and which may have implications for the health of certain sections of the population or give rise to ethical concerns. Examples would include allergenic potential or the use of animal genes, which may be the cause of religious concerns.
- The presence of an organism genetically modified by techniques of genetic modification as defined by MAFF guidelines.

The Food Labelling (Amendment) Regulations 1999 came into force in March 1999 in the UK and require that catering outlets, without exception, provide information concerning GM ingredients for their customers. The regulations apply to food and food ingredients produced in whole or part from GM soya or maize, but exclude foods where neither the protein nor DNA resulting from the genetic modification is present. In these cases the food is deemed indistinguishable from products produced using conventional soya or maize. Food additives, flavourings for use in foodstuffs or extraction solvents are also exempt as are ingredients lawfully sold before 1 September 1999. Similarly, products placed on the market before 1 March 1999, where other forms of wording (complying with EC Regulation 1813/97) have been used to indicate the presence of GM material, are also exempt. Further details can be obtained from Guidance Notes: Novel Foods and Novel Food Ingredients Legislation, and Guidance Notes: Labelling of Food Containing Genetically Modified Soya and Maize, MAFF.

7.5 Public perceptions

Opponents of genetic engineering are concerned that the technology is unproven and that not enough is known about the potential risks to human health or the environment. They argue that once genetically modified plants are released into the environment it will be very difficult to repair any damage caused by their release, and that humans are being used in a live global experiment to evaluate the safety of GM ingredients (Hill 1998).

Proponents of genetic engineering argue that the application of genetic engineering is controlled by strict regulations concerning the release of GM varieties into the environment. The regulations are concerned with ensuring that deliberate release, either for experimental or commercial purposes, does not pose a risk to the environment or to human health (Burrows 1999).

7.5.1   American versus European response

Despite the complex evaluations that a GM variety must pass before commercialisation becomes a possibility, coupled with transparency of the procedures, the British public has become increasingly worried by the development of these new varieties. Greenpeace commissioned a MORI poll in June 1999 to evaluate public opinion regarding the potential risk of gene transfer from GM crops to neighbouring organic crops (Anon. 1999). Mori approached 501 adults over the age of 16 and asked the following question:

> As it is currently agreed that some cross-pollination of GM crops with neighbouring organic crops is inevitable, it has been proposed that the standards for defining 'organic' foods should include organic crops which may have cross-pollinated with GM crops. How concerned, if at all, would you be if the definition of organic crops was changed in this way?

The results of the poll indicated that 45% of those questioned were very concerned, 29% fairly concerned, 15% were not very concerned, 8% were not at all concerned while 3% either did not know or had no opinion. The concern of the British public is in contrast to the attitude of North American consumers who have, until now, largely accepted GM foods.

The reasons for the difference in attitude between American consumers and European, in particular British, consumers have been debated long and hard, with the conclusion that differences in public perceptions are likely to be attributable to a number of causes.

- *Trust in Government and statutory bodies.*   A catalogue of poorly handled crises in the British food industry, ranging from the salmonella scare in eggs of the mid-1980s through to the recent BSE crisis, have left the public sceptical and mistrusting of official attempts to reassure it about the safety of GM technology. This distrust is in spite of the clear and significant differences in the actual risks associated with GM foods compared with these earlier incidents. In contrast, the USA has not experienced this crisis of confidence in the regulatory authorities and when the FDA move to reassure the public they appear to be successful.
- *Awareness of GM foods and what genetic modification involves.*   Recently it has started to become clear that the British public are considerably more aware of what genetically modified food is than their American counterparts. When consumers in the USA were interviewed at supermarkets to find out how aware they were of what GM foods are and how they regarded them, a significant proportion of those interviewed did not know what a GM food was and had not considered whether they may be desirable as part of their daily diet. The British environmental groups are beginning to report an upsurge of public concern in the USA in response to the furore seen in the UK and the rest of Europe.
- *Differing attitudes to the rural environment.*   There are clear differences in attitude to the countryside in the USA principally due to the way food crops

are farmed. While crop production in the UK and the rest of Europe forms an intrinsic part of our countryside and indeed has been largely responsible for creating the present look and feel of our countryside, crop production in the USA tends to occur at arm's length from the consumer in large acreages of land devoted to nothing else but highly intensive agriculture. In that sense there is a perception that the risk of gene transfer and environmental impact of GM crops on American wildlife is minimised and almost preventable whilst in the UK agricultural system the risk of environmental impact would be difficult to contain should a problem arise.

• *Lack of consumer choice.*   Of all the reasons for the current rejection of GM technology by the public, the lack of consumer choice is perhaps the most important. The decision to import GM and non-GM soya without segregation may well be taken as the turning point in consumer attitude against these new crops. Recognition that consumers demand the right to choose whether they eat GM foods has orchestrated a remarkable volte face by the large-scale food processors and supermarkets. They have moved almost unanimously to remove GM ingredients from their products. The knock-on effects can be seen in the elevated prices of last season's non-GM cereal and legume crops which commanded a premium.

New food labelling regulations came into force in the UK in 1999 and are aimed at meeting the demand for consumer choice. They require that catering outlets display information indicating that a product does or does not contain GM ingredients, with a maximum penalty of £5,000 for non-compliance. However, it is difficult to see how the scheme will work in practice. Given the technical demands of analysing and identifying GM ingredients, it will prove difficult for the local authorities to enforce.

The scientific community and industry are beginning to recognise that one of the problems in obtaining public acceptance of GM crops is the lack of benefit that these crops currently offer to the consumer. The development of GM crops has so far concentrated on producing resistance-enhanced crops. Herbicide, disease and insect resistances are the most prolifically produced GM crops world-wide. The benefits of these new traits are realised by the primary producers, the farmers, who do not have to spend money and time spraying to prevent crop damage. In the second major category of GM crop development, the emphasis continues to be on the added-value traits such as novel oil production. The benefits are enjoyed by processors rather than consumers.

Finally, in the smallest category are the designer traits such as improved nutritional value, shelf life or reduced cost, which is immediately obvious to the consumer. Even within this category the question of whether trait improvements such as slow ripening can be seen, or effectively sold, as a benefit to the consumer, rather than the wholesalers and supermarkets, is questionable. This lack of demonstrable benefits of genetic engineering to the public does little to diminish the suspicions that the real winners are the biotechnology companies, growers and processors.

7.6   Future developments in the regulatory process

7.6.1   Post-commercialisation regulatory systems

In view of the current upsurge of anti-GM feeling the regulatory controls surrounding the deliberate release of GM crops is likely to become even more stringent. Current regulations are under review with the aim of improving efficiency of the current procedures (Burrows 1999). Proposed changes to Directive 90/220/EEC include:

- Setting clear time limits during consideration of marketing release applications with the aim of giving a decision concerning an application within one year.
- Clarification of the risk assessment and harmonisation across member states to include assessment of direct, indirect, immediate and delayed risk or impact.
- Monitoring and time limitations for marketing consents, such that monitoring will occur post-release for commercial purposes. In addition marketing consents will have a finite life of ten years after which they must be reviewed.
- Improving the transparency of applications for consent to release for marketing, allowing the public to see the content of such applications.
- Appointment of a committee to deal specifically with the question of ethical issues arising from the application of biotechnology.

Risk assessments conducted for regulatory purposes tend to concentrate on the direct effects of the GM crop and its relatives on the natural environment. However, it is now becoming apparent that the agricultural consequences of the deployment and management of GM crops could also have significant impacts on the environment in regions where a very high proportion of the total land area is actively managed. Plans are being developed for monitoring the early years of the commercialisation of each GM crop so that its impact on both agriculture and the environment can be evaluated in farm-scale releases. The aims of such large-scale monitoring programmes are to:

- evaluate any different or unforeseen risks that might be associated with large-scale releases
- try to provide mechanisms of quantifying such risks
- try to understand the interactions of GM traits, for example herbicide tolerance, within the context of standard agricultural practice, e.g. crop rotations, over a longer period
- develop safe codes of practice to minimise risk.

There are a number of potential approaches to monitoring the post-release effects of GM crops. At this juncture in time when no GM variety has been released for commercial growth in the UK, and when the likelihood of imminent release is low, the priority must be to monitor experimental work.

7.6.2   Stewarding and voluntary codes of practice

As a response to customer concerns the UK agricultural and biotechnology industries have come together to form a stewardship group with its own voluntary codes of practice in order to develop an industry-wide code of practice for the commercial introduction of GM crops and their subsequent management. The group, known as the Supply Chain Initiative for Modified Agricultural Crops (SCIMAC) represents farmers, the seed trade, plant breeders and the agrochemical and biotechnology companies and was launched in June 1998.

SCIMAC seeks to provide traceability for individual consignments of GM varieties and outlines best practice for the provision of information at successive points along the food chain. Traceability begins with the plant breeders who must comply with EC regulations concerning the release of GM varieties into the environment for commercial purposes as discussed previously. SCIMAC then encourages seed merchants and growers to label clearly seed of GM varieties and to advise their respective customers of specific requirements concerning the growth, management and handling of these novel crops. Farmers must isolate GM crops from similar non-GM crops to prevent hybridisation and must keep all harvested products separate from non-GM crops. Merchants and wholesalers are required to maintain records concerning the storage and subsequent distribution off the farm. Any GM consignment to leave the farm must include a post-harvest declaration stating the variety name.

7.6.3   Food labelling

At the moment the most likely developments in the regulatory procedure are concerned with further modifications to food labelling requirements. The UK competent authority is expected to continue lobbying for the inclusion of food additives, such as lecithin, under the novel food regulations.

7.6.4   Multiple transgenes and risk assessment

The development of transgenic crops containing more than one transgene is already well under way, with a number of oilseed rape varieties containing herbicide tolerance genes and transgenes creating male sterility and restoring male fertility in hybrid varieties. These multiple transgene GM crops further complicate the risk assessment and may influence the decision process when considering applications for consent to release into the environment.

7.7   References

ANON. (1999) Greenpeace press release June 1999.
BIRCH A N E, GEOGHEGAN I E, MAJERUS M, McNICOL J W, HACKETT C, GATEHOUSE A M R and GATEHOUSE J A (1998) Ecological impact on predatory two-spot

ladybirds of transgenic potatoes expressing snowdrop lectin for aphid resistance, *Journal of Molecular Breeding*.

BURROWS P (1999) Deliberate release of genetically modified organisms: the UK regulatory framework. In 1999 BCPC Symposium Proceedings No 72: Gene flow and agriculture: relevance for transgenic crops, 13–21.

DANIELS R E and SHEAIL J (1999) Genetic pollution: concepts, concerns and transgenic crops. In 1999 BCPC Symposium Proceedings No 72: Gene flow and agriculture: relevance for transgenic crops, 65–72.

EWEN S W and PUSZTAI A (1999) Effect of diets containing genetically-modified potatoes expressing *Galanthus nivalis* lectin on rat small intestine, *The Lancet*, 354 (9187).

FAO/WHO (1996) Joint FAO/WHO Expert consultation on Biotechnology and Food Safety, Rome: FAO.

FENTON B, STANLEY K, FENTON S and BOLTON-SMITH (1999) Differential binding of the insectidal lectin GNA to human blood cells, *The Lancet*, 354 (9187).

GATEHOUSE A M R, DAVISON G M, NEWELL C A, MERRYWEATHER A, HAMILTON W D O, BURGESS E P J, GILBERT R J C and GATEHOUSE J A (1997) Transgenic potato plants with enhanced resistance to the tomato moth, *Lacanobia oleracea: growth room trials. Molecular Breeding*, **3**, 49–63.

GATEHOUSE A M R, BROWN D P, WILKINSON H S, DOWN R E, FORD L, GATEHOUSE J A, BELL H A and EDWARDS J P (1998) The use of transgenic plants for the control of insect pests. BCPC Symposium Proceedings No 71: biotechnology in crop protection: facts and fallacies.

HARDING K and HARRIS P S (1997) Risk assessment of the release of genetically modified plants: a review, *Agro-Food-Indust Hi-Tech*, Nov/Dec, 8–13.

HILDER V A, GATEHOUSE A M R, SHERMAN S E, BARKEER R F and BOULTER D (1987) A novel mechanism for insect resistance engineered into tobacco. *Nature*, **330**, 160–3.

HILL J E (1998) Public concerns over the use of transgenic plants in the protection of crops from pests and diseases and government responses. 1998 BCPC Symposium Proceedings No 71: Biotechnology in crop protection: facts and fallacies, 57–66.

HODGSON J (1999) Monarch Bt-corn paper questioned. *Nature Biotechnology*, **17** (7), 627.

KUIPER H A, NOTEBORN H P and PEIJNENBURG A C (1999) Adequacy of methods of testing the safety of genetically-modified foods, *The Lancet*, 354 (9187).

LOSEY J E, RAYOR L S and CARTER M E (1999) Transgenic pollen harm monarch larvae. *Nature*, **399**, 214.

MOYES C L and DALE P J (1999) Organic farming and gene transfer from genetically modified crops. MAFF research report.

NICKSON T E and McKEE M J (1998) Ecological aspects of genetically modified crops. Agricultural biotechnology and environmental quality; gene escape and pest resistance. NABC report no. 10, pp. 95–104.

NOTEBORN H P, BIENENMANN-PLOUM M E and VAN DEN BERG J H (1995) Safety assessment of the *Bacillus thuringiensis* insecticidal crystal protein CryIA(b) expressed in transgenic tomatoes. In Engel K H *et al. Genetically-modified foods: safety issues.* ACS Symposium Series 605, Washington DC: 134–47.

NOTEBORN H P, LOMMEN A, VAN DER JAGT R C, WESEMAN J M and KUIPER H A (2000) Chemical fingerprinting for the evaluation of unintended secondary metabolic changes in transgenic food crops, *J. Biotechnol* (in press).

OECD (1993) *Safety evaluation of foods derived by modern biotechnology: concepts and principles.* Paris: OECD.

OECD (1996) *Food safety evaluation.* Paris: OECD.

QIAN C M, XU A and LIANG G (1986) Effects of low temperatures and genotypes on pollen development in wheat. *Crop Science,* **26**, 43–6.

RAMSAY G, THOMPSON C E, NEILSON S and MACKAY G R (1999) Honey bees as vectors of GM oilseed rape pollen. 1999 BCPC Symposium Proceedings No 72: Gene flow and agriculture: relevance for transgenic crops, 209–14.

ROYAL SOCIETY FOR THE PROTECTION OF BIRDS (1997) The potential effects of releasing genetically modified organisms into the environment. RSPB.

SCHULER T H, POTTING R P J, DENHOLM I and POPPY G M (1999) Parasitoid behaviour and *Bt* plants. *Nature,* **400**, 825–6.

SENIOR I J and DALE P J (1996) Plant transgene silencing – gremlin or gift? *Chemistry and Industry,* 19 August, 604–8.

VAN HAL H N, VORST O and VAN HOUWELINGEN A M (2000) The application of DNA micro-assays in gene expression analysis, *J. Biotechnol* (in press).

WHITTON J, WOLFE D E, ARIA D M, SNOW A A and REISBERG L H (1997) The persistence of cultivar alleles in wild populations of sunflowers five generations after hybridisation. *Theoretical and Applied Genetics,* **95**, 33–40.

8

Current practice in milling and baking

A. Lynn, Scottish Agricultural College, Auchincruive

8.1 Introduction

The title for this chapter is deceptively simple. However, in describing the necessary stages and processes, the level of complexity and detail for each of the processes becomes apparent. The aim of this chapter is not to provide a comprehensive technical manual of each of the processing techniques used or subject areas covered, although some processing information is included. Instead the aim is to give the biotechnologist a basic understanding of the industrially important properties that cereals possess, and to try to highlight some of the properties that are considered important when processing them using current practices. For reasons that will be discussed, wheat will be the cereal that is referred to most frequently. However, other cereals (barley, rice, maize, oats and rye) will be considered where appropriate.

There is a very wide range of baked products that can be produced from cereals, although in one chapter it is impractical to attempt to cover all of the possible products. Instead this chapter concentrates on the properties of cereals that allow the production of breads and biscuits by the technologies employed in the Western world. This chapter also tries to emphasise the aims of milling and baking. At first this may seem obvious; the physical aims of milling are simply to fractionate the feedstock material into high- and low-value products, the value of the products being determined both by their usefulness and their availability in the market-place. However, commercial milling has other aims too. Commercial milling must be cost effective and efficient. It must try always to make the best compromises between efficiency, yields and costs. Instead of adding value to a product, it is very easy to add cost. That is to say, additional processes can be introduced, but if there is no commercial payback, then all that has been achieved is a reduction in the profitability of the process. In addition to commercial and processing considerations, the milling industry must always strive to meet the needs and demands of its customers, and ultimately the consumer of the finished products.

Few of the products of milling are consumed by humans without further processing. Baking is a particularly common process used to increase the nutritional value and acceptability of processed cereals. This chapter will discuss the production of bread and biscuits from cereals, although there are many other products that can be produced from milled cereals. The end goal of the baking process is to produce a product that has the physical and organoleptic properties that meet the specifications of the intended product. This aim seems relatively modest, but in practice it is quite a feat. In the mass-produced food market, food products are branded, and the consumer has learned to expect particular attributes to be associated with particular brands. Brands, however, tend to be associated with products that are widely available. This means that not all of the final product was produced in the same factory, or from the same raw materials, or even at the same time of year. However, the final product must comply with the characteristics of the branded food. Commercial baking has similar imperatives to commercial milling; it must be cost effective and efficient. It must try always to make the best compromises between efficiency, yields and costs. Ultimately, commercial baking must meet the needs and demands of its customers, i.e. the consumer.

8.2   Composition of cereals

The composition of cereals varies. Factors that influence the composition of cereals include variety, growth conditions, husbandry, disease and infestation. Approximately 65–85% by weight of whole grain (including hull if present) wheat, rice, barley, oats and maize is comprised of the starchy endosperm (Kent and Evers 1994). The remainder of each type of grain is bran and embryo. The economic imperative driving milling is to separate the components of the grains as efficiently as possible, and at the same time with minimal expenditure. The starchy endosperm of cereals is of most importance, as it contains proteins and starches that can be used for many purposes.

For some uses of cereals the starch in the starchy endosperm is the desired component of the grain, for other uses it is the endosperm as a whole that is important. The conditions of milling will be designed to maximise yield of the desired component, while imparting desirable characteristics (where possible, e.g. starch damage), and at the same time limiting the extent of any deleterious effects. In the case of starch extraction, the milling conditions will be adjusted so that the extracted material performs best in the subsequent separation steps used in the process. This often means that the milling of the cereal for this use is different from milling for other uses.

8.2.1   Starch biology

Starch is the name given to a plant storage carbohydrate. Plants deposit this material in various tissues, and for varying intervals of time. Some forms are rapidly metabolised, while others may persist for many days or even years. In most tissues the starch is deposited in the form of granules. The extent of starch granule diversity is large (Jane *et al.* 1994), although granules frequently have characteristics that allow the identification of the species which produced them. The starch granules of wheat, rye and barley contain two different populations of starch granules. The larger granules are called A-type granules, the B-type granules are the smaller granules. The granule categories have been further extended using some particle size determining techniques, although such categorisation has tended to be of academic rather than industrial concern.

The carbohydrates that are deposited to form the starch granules are the glucose polymers amylose and amylopectin. Both of these are high molecular weight polymers composed of α-glucan chains. Amylose is essentially linear (usually containing only a limited number of α-$(1\rightarrow6)$ branch points), while amylopectin is highly branched, the branches taking the form of α-$(1\rightarrow6)$ linkages. The exact structure of each polymer cannot be given, as each starch contains a population of carbohydrate structures. Some sources of starch contain a range of polymers that blur the distinction between amylose and amylopectin; material in this category is called the intermediate fraction.

Starch granules are not composed entirely of carbohydrate. Depending on the origin of the starch granules, they may contain varying amounts of lipids and minerals. The amount of the non-carbohydrate components of starch granules can be influenced by growing and extraction conditions that have been used to obtain the starch. These other components may have a significant effect on some applications for the starch, as they may influence, e.g. swelling and carbohydrate leaching properties. The swelling and leaching properties of starch can be very important in some applications. The swelling properties may determine the amount of water that is held in a starch-containing system. This could impact on the commercial viability of processing (beneficially and detrimentally), and also could affect other properties of the system, such as viscosity. Leaching can also affect viscosity. Materials that leach from the starch are able to interact with each other, and the other materials present, causing changes in technical properties such as viscosity.

8.2.2   Importance of starch damage to milling and baking

In the case of the milling of wheat to produce flour suitable for bread production, the process of milling causes changes to some of the properties of some of the wheat starch granules. Collectively these changes to starch properties have come to be known as 'starch damage'. During the purification of the starchy endosperm from the other components of the grain, some of the mechanical energy imparted to the endosperm is transmitted to the starch granules, where it causes structural changes to some of them. It is these structural changes caused by mechanical means that are called starch damage. It is clear that the starch granules and their interaction with the endosperm matrix, and the mechanical energy transmission properties of the whole grain, together with the milling process, will control the amount of mechanical damage that occurs to the starch granules.

It has been pointed out that the term 'starch damage' is misleading, as it implies a deleterious effect (Evers and Stevens 1985). To the miller the level of starch damage is not important (except to meet customer specifications), and a wide range of starch damage levels can be obtained by varying the mill settings. However, starch damage is commercially beneficial to the bread-baking process providing the level is not excessive.

The commercial significance of starch damage comes from the properties that mechanically modified starch granules possess. The commercially beneficial effects arise from the increased water-holding ability of damaged starch granules; affected granules swell more rapidly, and to a greater extent than 'undamaged' starch granules. Bread is often sold on a weight basis as opposed to a volume basis, therefore starch damage can increase the yield of bread from a given amount of flour. However, for a given flour there is a threshold value for starch damage above which increased starch damage causes bread quality problems, ranging from poor volume and crumb structure, dark crust, sticky crumb (causing slicing problems), perhaps even liquid loaf interiors. These problems are not necessarily purely the result of excessive water absorption, but also due to extensive α-amylolysis of the damaged starch granules, which are easier to digest than their 'undamaged' counterparts. Some of the problems may seem minor. However, industrial-scale baking and the use of continuous tunnel ovens means that bread is 'flowing' towards the slicing machines. If the slicing machines rip the bread instead of cutting it, due to stickiness of the crumb, then there is a big problem as bread is still being produced although there may be nowhere to put it or process it until the slicing machines are cleared.

8.2.3   Starch biochemistry

The biosynthesis of starch has attracted a great deal of interest over the last few years. Modern molecular biology techniques (Chapter 6) offer the potential to improve our understanding of starch biosynthetic processes. Similar techniques will allow the transformation of crops to produce commodities with desired characteristics (Chapters 2–5). Mutants that produce starches with different compositions have been produced commercially on a large scale for a number of years. The commercially grown mutants have tended to be those based upon the alteration of the relative proportions of amylose and amylopectin. Waxy starches are nearly all (depending on species) amylopectin, while so-called high amylose starches are composed of up to approximately 70% amylose, although 'high amylose' is lower in some species.

The following is a basic overview of a complex area which has been reviewed recently (Ellis *et al.* 1998). Mutant plants have aided the understanding of starch biosynthesis. It appears that there are three different classes of enzymes involved in the biosynthesis of starch, and therefore amylose and amylopectin. One class of starch biosynthetic enzymes is the starch synthases. These enzymes elongate carbohydrate molecules. A second class of starch biosynthetic enzymes called starch branching enzymes incorporate α-$(1\rightarrow6)$ linkages into carbohydrate molecules. A third class of starch biosynthetic enzymes are the debranching enzymes. These enzymes remove some branches of existing carbohydrate molecules and it is these prunings that are thought to act as primers for some starch synthase isoenzymes. Each of the classes of enzymes has been found to contain subgroups. The starch synthases have been subdivided into soluble starch synthases (three isoenzymes, SSS I, II and III) and granule-bound starch synthases (two isoenzymes, GBSS I and II). It is GBSS I and GBSS II that are thought to be responsible for the formation of amylose, while the action of SSS I, SSS II and SSS III in conjunction with starch branching enzyme (two isoenzymes, SBE I and II) is thought to be responsible for the production of amylopectin.

In the future, the use of molecular biology will be directed towards the production of amylose and amylopectin that is more tailored towards the end use of the starch. Specifically, it is likely that more control will be developed towards the production of starches containing amylose and amylopectin with more tightly defined structures. For example, the degree of branching of the starches, together with the chain lengths of the carbohydrate polymers could be more tightly controlled. However, there would still be some variation in these structures due to the environment in which the crop plant was grown. Bespoke starches could find markets in the food industry where starches are used extensively. Changes in the chain lengths or degrees of branching can cause differences in gelling, pasting and viscosity properties of starches. Bespoke starches would offer the cost saving associated with the omission of chemical modification processes that starches often undergo to induce desirable processing traits. The commercial uptake of such starches would be dependent upon the problems that occur as a result of raw material properties in some products, the cost of these problems, any problems with the bespoke starches themselves, and the amount of money that could be saved by solving the existing problems.

Where biotechnology could bring about radical changes in cereal starches is in the inclusion of properties in these starches that previously only existed in starches from other species. For example, potato starch naturally contains phosphate ester groups. These groups are useful in paper manufacture as they allow the starch to be attracted to the fibres, although this attraction is small in comparison to that achieved by some chemically modified starches (Laleg and Pikulik 1993). Gene transfer from potatoes to cereals of the necessary gene sequence(s) could open up more widespread use of wheat starch for paper production, or the use of cereal starches for novel applications. Both wheat and potato starches are currently used for paper and board production, but are often chemically modified (at a cost) to encourage binding of the starch. If the amount of chemical modification could be reduced, then this would have beneficial cost implications, and this would encourage the uptake of such starches. Other potential modifications depend upon the identification of genes that allow starches to be produced with similar properties to the chemically modified starches already produced. Starch modification *in planta* clearly has cost reduction implications, although such potential savings will be pursued only if the resulting starches are acceptable or advantageous, and the same price as (or cheaper than) existing starches.

8.3   Use of cereals in milling

Milling is a technological process that has developed as a result of the human ability to use tools to achieve particular aims. The aim of milling is to separate the components of the raw material. The technology which can be used to achieve this can range from hand-held implements (stick and stones), to vast arrays of machines. Each of these technologies has been adapted to the prevailing local conditions, be they social, economic, or agronomic. This next section cannot attempt to cover all the intricacies of milling; it will merely break the surface.

8.3.1   Scale and scope of milling

The milling industry is very large; world-wide grain production in 1997/98 was approximately 1.85 billion tonnes (Anon. 1999a). Of the grain produced each year, many millions of tonnes are milled. Each cereal has different milling characteristics, and within a type of cereal there may be further subdivisions of milling characteristics. There are therefore many different products of milling. This range of products is further expanded by the use of different grists. The grist is the mixture of different grains that forms the mill feedstock. A miller will vary the grist so as to produce a flour that meets the specifications of his customer. This means that a modern wheat mill will normally be able to produce tens of different flours for its customers, even though it has far fewer feedstock wheats. In addition to the natural variability, there is technological variation too. Not all cereals are milled to produce flours. Some cereals are milled to remove some or all of the structures that cover the endosperm of the cereal, prime examples of this being rice, oats and barley. In the case of rice, the whole grain is passed through a 'sheller' which removes the hull from the grain. After shelling, the rice retains the bran layers physically attached to the endosperm; this is brown rice. The brown rice grains are then abraded by rubbing them together under pressure. The abrasion removes the bran layers to produce white or polished rice. The milled material can then either be processed into a food material, or can undergo further milling stages to produce a flour if required.

8.3.2 Mill types and technology

Milling is a mechanical process. As previously mentioned there can be many mechanical technologies applied to solve particular problems. In the Western world roller milling of cereals is the most common method used, and the following discussion is based on this method of wheat milling.

Cleaning

The first stage the wheat undergoes before milling to produce flour is cleaning. It is essential that as much of the contaminating material as possible is removed from a sample of wheat. The wheat will be separated by size on sieves to remove much larger and smaller objects, by shape using indented discs or cylinders to remove other grains present, and specific gravity to remove stones. The grain is exposed to air currents to help remove minor contaminating matter, and undergoes scouring, which removes dirt from the exterior of the grain, including material lodged in the ventral crease.

Conditioning

If the wheat were to be milled immediately after cleaning, then the flour produced would not be of the highest grade. This is not because it would be in some way unwholesome, but rather because the milling of grain in this fashion would produce a flour that was contaminated with a significant amount of bran material. This bran material would affect the colour of the flour, and one of the miller's prime objectives is to separate the components of the grain effectively, i.e. to produce as clean (white) a flour as possible. To reduce the amount of bran particles that contaminate the flour, wheat is conditioned before it is milled. Conditioning involves the controlled addition of water to the wheat prior to its milling. After water addition, the wheat is held for a period of time to allow hydration of the wheat to occur. The addition of water during conditioning of the wheat makes the bran less friable (less likely to pulverise, and crumble) during milling.

Different wheats require different levels of conditioning; generally North American wheats are conditioned to ~1% higher moisture content than that of UK wheat (15–16% moisture). Addition of too much water can be problematical, as the products of milling will not flow through the machinery efficiently and may cause clogging of sieves, or blockages in pipes. Automated conditioning is common in modern flour mills.

Milling

Once the wheat has been correctly conditioned, it is milled. There are a large number of variations, both in terms of exact process specifications (many different systems are used), and the number of options available after each milling and separation stage. It is therefore difficult to give an exact path that material would take through a mill, as the processing of material in a mill is based upon its processing requirements, rather than on a predetermined pathway (although there are, obviously, a finite number of processing stages).

The first stage in wheat milling is performed by break rolls. Break rolls comprise a pair of counter-rotating metal rollers. These rollers are helically fluted or corrugated so that they can 'hold' the grain as it is milled. The flutes or corrugations on the break rollers are important, and cannot be adjusted (replacement or redressing is the only way to change the fluting). Their design is selected specifically for the type of material that is to be milled. There is a speed differential between the two rollers; one of the pair rotates about 2.5 times faster than the other roller. This speed differential is important, as grains that are held in the flute of one of the rollers are opened up by the differential movement of a flute on the other roller. Break rolls are designed to open up the grains. They do not produce much flour, although any flour that is produced at this stage has low bran content making it appear very white. This fraction also has an elevated protein level compared with other flour fractions. For these reasons, 'first break flour' can command a price premium.

After each milling stage, the material that has been milled is separated into a number of fractions. Each fraction is processed differently. The break rolls should produce large fragments of endosperm, which can then be further milled by reduction rollers.

Reduction rolls are the means by which large endosperm chunks are converted into flour. There are a series of reduction rolls, so that endosperm chunks are not resized in a single step. If this were done, high temperatures that could affect gluten properties would be produced. The severity of a one-step reduction would also pulverise bran particles, making them harder to separate. The flour produced would therefore be of a lower grade owing to its darker colour. Reduction rolls are smooth (except for a slight surface texture), and do not contain flutes. Reduction rolls, like break rolls, are counter-rotating, although reduction rolls operate at a much lower speed differential, approximately 1.25:1. Some of the mechanical energy that rolls deliver is converted into heat energy. To avoid heat damage to the flour, rolls (including break rolls) are cooled. Some mechanical starch damage is desirable in a bread-making flour, and it is during the reduction milling of the wheat that starch damage is generated as a consequence of the mechanical energy transfer through endosperm pieces.

After each milling process the products of milling are categorised, and similar material mixed together and processed according to its needs. Plansifters contain sieving surfaces mounted within a housing. The whole plansifter body moves in a rotary motion around its vertical axis. This motion distributes the plansifter feed material over the separating surfaces. The size of the holes in the sieving surfaces is selected to be appropriate for the feed material of the plansifter. Material retained by each of the sieving surfaces is separated into separate outflows, and is then processed by the appropriate milling stage.

Purifiers can be used to screen the material separated by the plansifters (if white flour is to be produced). This screening is necessary as the outflow from plansifters can contain bran fragments of a similar size to endosperm fragments (therefore not separated by a plansifter). Purifiers separate this bran, and also fractionate plansifter outflow semolina (fragments of starchy endosperm) based on size and degree of bran attachment. This fractionated material is then sent to the appropriate break or reduction roll stage.

Following each plansifting stage, there may be a fraction that could be included in the final flour. The different flour streams are combined in the correct ratio to produce a flour with the required characteristics. This applies to wholemeal and brown flours too, as it is often more economical to mill these flours as white, and then to recombine the different fractions in the correct ratios.

Alternative technologies
While the above system is very common, it is not the only system for producing wheat flours. New technologies are constantly being developed. Whereas the current technology opens up the grain, and basically removes as much endosperm from a sheet of bran as possible, some new equipment takes a different approach (Anon. 1999b). Such machines keep the grain intact as far as possible. The first stage is the removal of the bran layers from the wheat through a series of abrasion steps, followed by polishing of the endosperm. The endosperm can then be hydrated before being milled. The claimed benefits of this approach include higher extraction and better quality flours, together with a reduction in the other milling stages required and lower power requirement.

Wet milling is an alternative milling technology. Wet milling is often used if the aim of the milling is not to separate the components of the grain into a flour and bran, but to separate all of the components of the grain, e.g. to separate the grain into bran, starch, protein, and, depending on the grain, oil. Wet milling is predominantly used on corn (maize) although it can be used to mill other cereals, e.g. wheat. The aim of wet milling of corn is to allow separation of the germs and oil-containing portions from the more fibrous starch and protein-containing material. The actual milling consists of a primary grinding stage which is performed on kernels that have been soaked in warm (50°C) acidified (to prevent excessive microbiological activity during soaking) water for between 24 and 48h, followed by separation of the germ material. The suspension is then further milled by either impact or impact-attrition milling. These techniques are based on the feed material hitting a fast rotating shaft that has bars or paddles protruding from it. The impact with the bars/paddles assists the separation of the starch from the protein and fibrous matter. The identified corn characteristics desirable for wet milling are as yet relatively non-specific. Corn that will wet mill to give a good starch yield is corn with large kernels, and the packing in the large kernels is soft. If corn is to be dry milled it should have large kernels, there should be a small amount of kernel size variation, and the kernel texture should be hard with an elevated protein content.

8.4   Cereal requirements for milling

Cereals to be used for milling must fulfil a number of requirements to ensure the products of milling are produced economically, and that they meet the requirements of the consumer. The following sections primarily describe the requirements for wheat, although equivalent considerations will be necessary for other cereals. General specifications for other cereals will include variety, grain size (preferably large and uniform), disease (absence), infestation (absence), damage (limit) and sprouted (absence/limit).

8.4.1 Moisture

Hydration of stored cereals is important, as prior to milling the moisture of cereals is adjusted to a level that allows optimal processing. Bran layers can be separated more effectively if the moisture level is optimal.

Cereal moisture levels are measured prior to acceptance of a cereal by a miller. The moisture level of cereals is critical for the economics of the milling process. If the moisture level is too high, then the milled cereal will not perform efficiently in the mill, and may cause blockages on sieves and in pipes due to aggregation of moist material. If such blockages occur, the downtime is expensive and problematical. A second problem caused by excessive moisture is that of storage. Routinely cereals are stored before being milled, and if they have excessive levels of moisture in them, they are prone to deterioration due to microbial activity and other infestations. Clearly materials with excessive levels of moisture cannot be stored, and if they are to be accepted by a miller, drying is necessary. This is expensive, and in a buyer-dominated market, affected cereals will either be rejected, or be bought for a reduced price.

8.4.2 Grain weight and density

Wheat samples are routinely tested for their thousand grain weight. The thousand grain weight is, as its name suggests, simply the weight of one thousand grains. It is a measure of grain size and quality. Heavier grains contain more endosperm, and therefore have the potential to yield more flour during milling.

A related test is the test weight, in which a fixed volume of grain is weighed. This test measures how well a sample of grain packs together, and the density of the grains. A low test weight is indicative of a low-density grain that packs poorly. This could occur as a result of grain roughness or unevenness, and grain with a mealy (loosely packed), or poorly filled endosperm. Conversely, a high test weight is indicative of a high-density grain which packs together well. Grain with a high test weight therefore has the best potential for a good yield of flour.

8.4.3 Wheat grain hardness

Grain hardness is genetically determined. Grain is split into two main categories, hard or soft. When hard wheat is milled, the flour produced is composed of particles that tend to be gritty or granular. This is beneficial as these particles are free flowing in the pneumatic and aspirated systems that are present in modern flour mills. Another beneficial effect of hardness derives from the mechanical energy transmission properties of the endosperm of hard milling wheat. Starch granules in a flour produced from a hard wheat are more likely to be mechanically damaged than starch granules in a flour produced from a similarly milled soft wheat. Hard wheat therefore carries certain economic benefits that make it preferable for producing flour for bread making (Section 8.2.2). One consequence of the difference in susceptibility of hard and soft wheats towards starch damage is that soft wheats are used for the production of flours where the products of baking are relatively low in moisture, e.g. biscuits and cookies.

8.4.4 α-amylase level

Wheat samples are tested for the level of endogenous α-amylase that they contain. These tests are important because of the problems that excessive α-amylase activity causes during baking. Grain with an excessive level of α-amylase will be rejected, as it is costly and difficult to reduce the α-amylase activity of a flour without having a deleterious effect on the desirable properties of the other grain proteins. If the α-amylase is slightly elevated, but not excessive, then the grain may be purchased but at a reduced price.

8.4.5   Protein

As protein content and quality of wheat is important for its end use, it is routine for the wheat to be tested for its protein content. Testing methods vary depending on requirements, and include Kjeldahl determinations, NIR, gluten washing and sedimentation testing.

8.4.6   Flour yield

Test milling of a small sample of wheat can be performed to predict how well a particular wheat will perform on the large-scale milling machinery.

8.5   Use of cereals in baking

Cereals are the staple food of much of the world, and have been for many thousands of years. As a result of the different climatic conditions in different regions of the globe, some cereals are more suited to growth in one region as opposed to another. Raw cereals are not the most palatable of foods and they do not yield maximal nutritional benefit from the cereal. Regional populations have therefore developed processing methods that allowed them to gain the most nutritional benefit possible from the available food sources. One of the most common methods of achieving this is to produce a bread. A number of cereals can be used to do this, although a number of the breads produced would not look like the breads available in the Western world. The reason for this is that there are many different ingredients used, and the role of the cereal is critical. Although it has been mentioned that a number of different cereals can be used, wheat is by far the best cereal for the production of leavened bread products, with good crumb structure, and large loaf volume. If the volume of the loaf is less critical, then other cereals can be used.

Of the available wheat in the world, not all of it is suitable for bread making, although technological advances have allowed more sources of wheat to be used for the production of bread. Good bread-making wheats are known to come from a number of regions around the world. The genetic makeup of the plants in combination with their growth conditions produces wheat which is high in protein (compared with other sources of wheat), and these proteins are high-quality proteins too, that is, they are able to be processed to produce good quality bread. Desirable property traits will include (among others) properties such as extensibility, elasticity and the ability to hold gas bubbles. The role of cereals in the production of bread and biscuits will now be examined in more detail.

8.6   Bread baking

There are many different types of bread produced all around the world. These breads are different as a result of differences in the cereals available in specific locations, and the processing technology that is applied to convert the cereals used into bread. Bread is a staple food of many millions of people, and it is eaten and produced in many different ways. Fundamentally, however, most breads can be considered as edible foams which are produced through the application of 'traditional biotechnology'. Bread can be produced from a number of cereals, although given a ready supply of any preferred cereal, wheat is the cereal that is best suited to producing loaves that expand due to fermentation to produce high volume breads. This suitability is a result of the properties of the wheat proteins (Skerrit 1998). The proteins responsible for the useful properties of wheat flour are gliadins and glutenins, which collectively are known as gluten or vital gluten. The elastic properties of these proteins are what makes wheat flour so good for bread making, and care should be taken during all processing to ensure that these properties are not adversely affected.

Cereals other than wheat can also be used for bread production. Rye bread has been a staple food of many of the world's people. It does not produce bread with as good a crumb structure or loaf volume as wheat bread, although it performs adequately. A small addition of wheat flour to rye flour can effect a large improvement in the baking performance of rye flours. Rye flour can contain elevated enzyme levels which are deleterious to bread production. This can be a result of premature germination of the rye in the field, before it has been harvested and milled.

Barley too can be used for bread production as it contains some gluten. It is not a mainstream bread cereal, although it can be added to wheat-flour grists in small amounts. Again, it does not produce bread with as good a crumb structure or loaf volume as wheat bread.

8.6.1 Baking processes

A baker cannot produce a good loaf from a poor flour. However, the baker can change processing conditions so as to optimise his product from a given flour. There are many different baking processes employed around the world. They range in scale from home baking, to large industrial-scale bakeries. In the UK the majority of bread is produced on a large industrial scale. As a result of the scale of the industry, baking has moved from a slightly haphazard process to a highly automated and controlled production process. This change has progressed at a higher rate in the latter half of the twentieth century with the implementation of mechanical dough development technology.

The traditional bread-making process involves the mixing of the bread ingredients, followed by long periods where the fermentation process is allowed to proceed. As the fermentation progresses, the activity of the yeast causes changes in the properties of the dough. For example, the gas cells produced during mixing begin to expand as a result of the carbon dioxide produced through yeast respiration. During fermentation the dough proteins become more 'relaxed' and extensible too.

In an industrial process it is desirable to avoid periods of inactivity for commercial reasons. In the 1950s it was found that the resting periods could be omitted if large amounts of mechanical energy were imparted to a dough via mixing. This idea was developed specifically for UK-produced flours at the Flour, Milling, Baking Research Association, at Chorleywood in Hertfordshire. The bread-making process developed there became known as the Chorleywood Bread Process or CBP for short. The CBP has a number of advantages over traditionally produced bread. Flour for the CBP can be lower in protein content than flour used for the traditional process. This is significant as more UK-produced flour can be used to make bread, as UK-produced wheat is naturally lower in protein than some other sources of wheat. The CBP process also has other advantages in terms of bread yield, consistency and crumb colour over the traditional process.

8.6.2 Chemistry of bread making

The protein network produced during baking forms the physical backbone of the bread, and hence is very important for loaf volume and crumb structure. Flours from wheat relatively low in protein (quantity and quality) are more suitable for low loaf volume products such as baguettes, whereas higher protein wheat flours are more versatile, and can be used to produce higher loaf volume products, as they contain sufficient protein to form extended support structures. The industrial bread-making process relies on the chemical properties of the gluten proteins. These proteins contain a number of features that allow leavened bread to be produced. The glutenin proteins contain structural features that confer bonding and elasticity properties to the molecule. The glutenin proteins are rich in glutamine, non-polar amino acids, and charged amino acids. These groups respectively allow hydrogen bonding, hydrophobic interactions and electrostatic interactions.

While these interactions are important, the most significant bonding between gluten proteins occurs through the formation of disulphide bonding. The disulphide bonds lock together protein subunits, and thus allow the formation of extended protein networks. The elasticity of gluten originates from the structure of the glutenin subunits. The subunits of glutenin can be classified as either high molecular weight (HMW) or low molecular weight (LMW) subunits, and there are many different proteins in each category. Variation in the HMW subunits has been found to have a direct impact on the bread-making quality of wheat. These subunits and the DNA sequences coding for them have been studied intensively over a number of years. The subunits have regions that are relatively inelastic, but contain thiol groups capable of forming disulphide bonds with other subunits. Sandwiched between these regions is a region that contains a repeating structure. This repeating structure contains secondary (repeated β-hairpin turns) and tertiary structures (β-spirals) that confer elasticity to the molecule. Hexaploid bread wheats contain six genes that code for HMW subunits; each expressed contributes about 2% of the total gluten proteins. However, between one and three of these genes is not expressed, and therefore the gluten of such wheats contains between 6 and 10% HMW subunits (Shewry et al. 1997). The LMW subunits of glutenin are thought to crosslink (via disulphide bonds) the protein network of gluten, while the gliadin molecules are mainly included through hydrogen and hydrophobic bonding, and thus they both contribute to the gluten matrix.

The industrial processing of wheat to produce bread relies on the mechanical energy of mixing to disrupt, rearrange and align the dough proteins. If the dough were to remain in that state it would not achieve maximum elasticity and extensibility as there would be no extensive protein network. Bread produced from such a dough would have relatively poor crumb structure and loaf volume. The natural reformation of disulphide bonds would lock the proteins into a new network structure. However, this would be a relatively slow process in comparison with the requirements of the industry. A number of chemical compounds have been used to speed up the reformation of disulphide bonds. One of the standard agents is ascorbic acid (vitamin C). In addition to chemical ingredients added to bread to aid production, a number of enzymes can be added to dough to bring about changes beneficial to bread production. For example, lipoxygenase is an enzyme found in soya flour. It has a brightening effect on the crumb colour, and encourages disulphide bond formation giving rise to larger volume loaves. Transglutaminases have been added to doughs, especially to prefermented doughs. They catalyse the acyltransfer reaction between the γ-carboxyamide group of peptide-bound glutamine residues and primary amines. Transglutaminases therefore catalyse the linking of protein molecules via the side chains of glutamine and lysine residues. This is therefore another method of stabilising the protein network of a dough. A number of other enzymes can also be added to aid with the processing of the bread. The use of enzymes in food products is controlled by regional legislation.

8.7   Biscuit manufacture

Biscuits (cookies) are entirely different products from bread, and as such the cereals that are used to produce them are required to meet different specifications. A range of cereals can be used for biscuit production. If wheat is used, it is often soft wheat that has a relatively low protein content when compared to a bread wheat. Biscuit wheats tend to produce more break flour than bread wheats, and the flour that is produced from the biscuit wheat usually has a smaller average particle size. The beneficial effects of starch damage referred to in Section 8.2.2 are not beneficial when considering the production of biscuit flour. Starch damage is not desirable in this instance, as it would increase the moisture content of the product. Bread is a relatively high moisture content product (35–45%), but biscuits are lower. Biscuits containing high levels of moisture would tend to have poorer organoleptic (sensory) properties than 'dry' biscuits, e.g. the 'snap' of the biscuit or its mouthfeel would not be correct. The proteins in biscuit flour have to form a network throughout the biscuit, although unlike bread proteins, the network does not have to be able to expand as much as bread, as the amount that biscuits increase in volume during baking is much lower than that of breads. The rheological properties of biscuit dough are also of importance. This is because biscuit dough is often rolled into a sheet, from which biscuits are cut. For biscuits, it is therefore a requirement that the proteins in the flour must be extensible (i.e. be able to be rolled into a sheet) but the sheeted dough should have minimal recovery of shape, i.e. it should not contract after being rolled. The rheological properties of biscuit doughs are particularly important in highly automated production facilities. If the dough were to contract after the biscuit pieces had been cut out, the biscuits would have a smaller diameter, and be thicker than expected. This variation would pose problems for the packaging machinery.

8.8   Summary

8.8.1   Role of biotechnology

Cereals form a significant part of the diet of the world's population, and the effective processing of cereals is therefore of global importance. Cereal processors have to try to meet the demands that are placed on them, both in terms of product composition, and commercial imperatives. One of the things that all processors of primary products must cope with is the degree of variability that can be experienced with raw materials. The variability can be significant, and this leads to processing problems that need to be resolved. Modern flour milling is a tremendous technological achievement. By a series of individually simple mechanical processes, a biological product with varying composition and dimensions can be processed into a product that complies with predefined standards. The degree of automation in a new flour mill means that only two or three actual millers need to be employed. They will monitor the process, taking corrective action if the process deviates from tolerances predefined in the controlling software. All that needs to be done manually is to key into the computer the amount of product to be produced, and the time at which it is required. The software controlling the mill will mix an appropriate grist (the test results for each batch of wheat are also in the computer) and process it correctly to ensure that the delivery schedule is met. All of this has been achieved by developing engineering and technology to address a biological problem. This was the route taken as it was faster to alter the processing of the cereal than to alter the cereal itself.

Modern biotechnological methods now, or will shortly, allow crop plants to be modified far more quickly and specifically than has previously been achievable. This means that in addition to varying the processing conditions, it will be practicable to vary the raw material. One of the targets that will be aimed at by the new biotechnologies will be to adapt the crops so that in addition to good agronomic, disease resistance and general handling properties, it also has good processing properties.

8.8.2   Breeding and selection problems

One of the problems that industry has so far experienced is that processability has not been high on the plant breeders' selection criteria. Indeed, some of the more modern varieties have proved to be more problematical to process than their predecessors. Processing has been hindered for example by excessive stickiness of some new crosses. Hitherto, traits such as disease resistance, yield or straw properties would be the selection criteria, and crosses would be screened against these traits. As certain lines had historically worked well in particular processes, these lines were improved and used for those specific purposes. Crosses which failed the screening would be discarded, and of those that passed, a small number would then be tested in specific applications (there are problems with producing sufficient amounts of material to be tested and in the length of time it takes to screen large numbers of samples). Of those that failed the screening there may have been individual crosses that outperformed any of the successful candidates in terms of processability, and may have given clues as to the genetic basis of this improved performance; however, these would have been discarded. Given that the genetic basis for some processing traits will be identifiable, the selection of wheat for particular industrial applications will be possible for each new cross. Therefore there should be an improvement in the efficiency of individual processes.

8.8.3   Impact of biotechnology

Genetic engineering of cereals for particular purposes could have any number of specific target applications, and the following examples are just scratching the surface of the alterations that may be thought of or attempted.

Enzymes

There are a number of instances where enzymes are added to cereals during processing; these could be incorporated into the cereals through genetic engineering. One use of cereals is the production of animal feeds. Animal digestion of cereal derived material can sometimes result in viscous gut contents caused by β-glucan. Animal feeds can have β-glucanase added to them to avoid the costly problems associated with viscous gut contents and faeces. These problems include slow animal growth (increasing costs), and quality down-grading of carcasses due to adhered faeces (reducing profits). Clearly, a plant with highly active β-glucanase in it would probably not grow satisfactorily; however, one solution would be to produce a modified and inactive β-glucanase *in planta*, which would then be activated by the animals' digestive processes.

Another possibility is that endogenous enzyme genes could be replaced with genes that coded for similar enzymes with different properties. For example, α-amylase genes could be exchanged so that the modified plant produced amylases that were less troublesome in industrial processes.

Improved separation

In the case of milling of cereals, it is clear that modifications that aid the separation processes would be welcomed. Modifications to the mechanical properties of the bran layers/endosperm interface that allowed the bran to detach itself from the endosperm more cleanly would be beneficial. Developments that would allow the alteration of the mechanical properties of bran layers would also be potentially beneficial, as in some separation processes friable bran is undesirable as it colours the product, whereas in other processes easily abraded bran layers would be beneficial too. If the friability of the bran layers could be controlled then conditioning of cereals could be reduced or even omitted. The industrial process would then have fewer steps, making it cheaper and faster to operate. Friable and easily broken cereals do not perform well in impact milling (used to remove the outer layers of cereals). If this could be improved, then processing would be faster and more economical.

Grain shape

Genetic modification to grain shape may have beneficial effects in some abrasion processes, by reducing the depth of the ventral crease, and thereby making the abrasion of the bran material easier.

Pigmentation

The pigment in the exterior of the wheat could be reduced, eliminated or altered. A lighter bran would mean that flours could look whiter, yet still contain bran.

Millers and bakers often like to produce white flours and breads as customers generally prefer them. 'White' 'brown bread' is probably not possible, but it may be possible to produce a white bread containing a significant amount of bran.

Granule size

When wheat or barley starch is being produced commercially a great deal of effort is spent separating large and small granules from the associated protein and bran material. If the genetic basis for the initiation of large and small starch granules could be understood, then it would be possible to produce cereals that had monomodal granule size distributions where they had previously been bimodal. Two crops could then be grown, large granule wheat and small granule wheat.

There are many more possibilities that can be thought of as to how biotechnology could be used to improve industrial processing of cereals. The excitement of the biotechnologist as a result of the multitude of potential changes that may be made, is matched only by the uncertainty and trepidation that the public has for the new technologies. This has been demonstrated by the numerous news headlines and scare stories that have been produced in the British press recently. In the introduction to this chapter it was stated that both industries must always strive to meet the needs and demands of their customers. It is well known in the field of marketing that if you want to sell something, you should emphasise benefits rather than features. If the biotechnology industries do not manage to convince the consumer of the benefits of genetically engineered crops, then the demand may be for the customer's needs to be met through traditional breeding techniques. The new biotechnologies would have an important role to play, but at the screening stage, as opposed to the genomic manipulation stage.

The future of cereal biotechnology will present many hurdles to overcome, some technical, some emotional and some rational. Cereal biotechnology will therefore definitely be interesting and exciting for the foreseeable future.

8.9 Bibliography

ANON. (1999). American Association of Cereal Chemists. [Online]. Available http://www.scisoc.org/aacc/ [1999, May 25].

ANON. (1999, March 30). Biscuit Equipment, Inc. [Online]. Available http://www.biscuitequipment.com/ [1999, May 25].

ANON. (1999). Biscuit Making. [Online]. Available http://www.crop.cri.nz/foodinfo/millbake/biscuit.htm [1999, May 25].

ANON. (1999). Federation of Bakers [Online]. Available http://www.bakers federation.org.uk [1999, May 25].

ANON. (1999). Food Resource. [Online]. Available http://osu.orst.edu/food-resource/index.html [1999, May 25].

ANON. (1999). Flour advisory bureau [Online]. Available http://www.fabflour. co.uk/ [1999, May 25].

ANON. (1999). National association of wheat growers home page. [Online]. Available http://www.wheatworld.org/ [1999, May 25].

ANON. (1999). Official documents home page. [Online]. Available http:// www.official-documents.co.uk/ [1999, May 25].

ANON. (1999). The corn refining process. [Online]. Available http://www.corn. org/web/process.htm [1999, May 25].

ANON. (1999). The maize page general information. [Online]. Available http:// www.ag.iastate.edu/departments/agronomy/general.html [1999, May 25].

ANON. (1999). The New Zealand Cyberguide to Flour Milling and Baking. [Online]. Available http://www.crop.cri.nz/foodinfo/millbake/home.htm [1999, May 25].

ANON. (1997, September). PBI Bulletin. [Online]. Available http:// www.pbi.nrc.ca/bulletin/sept97/sept97.html

ANON. (1999). Stoneground Mills. [Online]. Available http://www.infoweb.co.-za/stoneground/ [1999, May 25].

ANON. (1999). Recent News & Features. [Online]. Available http://www. sosland.com/content/recent.htm [1999, May 25].

ANON. (1999). USA Rice-Milling. [Online]. Available http://www.usarice. com/RICEINFO/milling.htm

ANON. (1998). Flour Based Foods. [Online]. Available http://www.awb.com. au/grain_ind/florfods.htm [1999, May 25].

ANON. (1998). USDA-NASS, Agricultural Statistics 1998. [Online]. Available http://www.usda.gov/nass/pubs/agr98/acro98.htm [1999, May 25].

ANON. (1998, March 24). World corn production. [Online]. Available http:// www.corn.org/web/wcornprd.htm [1999, May 25].

ATKINSON G (1999, May 18). Market Situation and Outlook. [Online]. Available http://www.agric.gov.ab.ca/economic/outlook/milling_wheat.html [1999, May 25].

CHENG L M, *Food Machinery for the Production of Cereal Foods, Snack Foods and Confectionery*, Woodhead Publishing, 1991.

CLOUTIER S and PENNER G A (1998). Cloning hmw and lmw glutenin genes from the extra strong bread wheat cultivar glenlea. [Online]. Available http://probe. nalusda.gov:8000/otherdocs/pg/pg5/abstracts/p-5c-158.html [1999, May 25].

COOPER J (1999). Flour mill. [Online]. Available http://www.rsf.co.uk/flourmill. htm [1999, May 25].

FIELD E K (1998, September 16). Post Harvest Grain Quality & Stored Product Protection Program Home Page. [Online]. Available http://pasture.ecn.purdue. edu/~grainlab/sites.html [1999, May 25].

FOX S L and HUCL P (1997, September). Wheat Gluten – more than just bread! – Western Canadian wheat breaders' perspective [Online]. Available http:// www.pbi.nrc.ca/bulletin/sept97/west.html [1999, May 25].

HENRY R and KETTLEWELL P (eds), *Cereal Grain Quality*, Chapman and Hall, 1995.

JACKSON D S (1996). Common industrial uses for corn. [Online]. Available http://foodsci.unl.edu/OnLineEd/Grain/cornqual/cqciu.htm http://www.ianr.unl.edu/pubs/FieldCrops/g1115.htm [1999, May 25].

JUKES D J (1999). Food Law. [Online]. Available http://www.fst.rdg.ac.uk/foodlaw/index.htm [1999, May 25].

MANLEY D J R, *Technology of Biscuits, Crackers and Cookies – 2nd edn*, Woodhead Publishing, 1991.

MURRAY MCLELLAND (1996, September 20). Wheat Utilization. [Online]. Available http://www.agric.gov.ab.ca/crops/wheat/wtmgt02.html [1999, May 25].

OWENS G (1997, August 9). Milling. [Online]. Available http://www.iol.ie/~gavo/gavpage.htm [1999, May 25].

PALMER G H (ed.), *Cereal Science and Technology*, Aberdeen University Press, 1989.

POSNER E S and HIBBS A N, *Wheat Flour Milling*, St. Paul, MN, AACC, 1998.

SCHNEEWEISS R, *Dictionary of Cereal Processing and Cereal Chemistry*, London, Elsevier, 1982.

SHELLENBERGER J A, in *Advances in Cereal Science and Technology III*, AACC, St Paul, MN, 1985.

SHELTON D R (1999). The Mill. [Online]. Available http://ianrwww.unl.edu/ianr/agronomy/wheatlab/mill.htm [1999, May 25].

SINGH N and ECKHOFF S R (1995, June 12). Hydrocyclone Operation for Starch-Protein Separation. [Online]. Available http://home.earthlink.net/~nsingh/publications/w-m-note.html [1999, May 25].

TIEDE M (1997). The structure of wheat gluten and its effect on dough making properties. [Online]. Available http://poa.ag.uidaho.edu/plsc501/f97/abstracts/mtiede_abstract.html [1999, May 25].

Note
Links to all the online material in the References and Bibliography sections can be found by following the links from http://www.foodfocus.freeserve.co.uk

8.10   References

ANON. (1999a). USDA-National Agricultural Statistics Service, Agricultural Statistics. [Online]. Available http://www.usda.gov/nass/pubs/agstats.htm [1999, May 25].

ANON. (1999b). The satake homepage. [Online]. Available http://satake-usa.com/ [1999, May 25].

ELLIS R P, COCHRANE M P, DALE M F B, DUFFUS C M, LYNN A, MORRISON I M, PRENTICE R D M, SWANSTON J S and TILLER S A, 'Starch production and industrial use', *Journal of the Science of Food and Agriculture*, 1998, **77**(3) 289–311.

EVERS A D and STEVENS D J, in *Advances in Cereal Science and Technology VII*,

AACC, St Paul MN, 1985.

JANE J L, KASEMSUWAN T, LEAS S, ZOBEL H and ROBYT J F, 'Anthology of starch granule morphology by scanning electron-microscopy', *Starch-starke*, 1994, **46**(4) 121–9.

KENT N L and EVERS A D, *Kent's Technology of Cereals. An Introduction for Students of Food Science and Agriculture. Fourth edition*, Woodhead Publishing, 1994.

LALEG M and PIKULIK I I, 'Modified starches for increasing paper strength', *Journal of Pulp and Paper Science*, 1993, **19** 248–55.

SHEWRY P, TATHAM A, BARCELO P and LAZZERI P (1997). Quality Improvement. [Online], Available http://www.pbi.nrc.ca/bulletin/sept97/genetic.html [1999, May 25].

SKERRIT J H (1998). Gluten proteins: genetics, structure and dough quality – a review. [Online], Available http://agbio.cabweb.org/REVIEWS/AUG98/HTML/SKERRITT.HTM [1999, May 25].

9

Current practice in malting, brewing and distilling

R. G. Anderson, Marchington Zymoscience, Uttoxeter

9.1 Introduction

Cereal-based alcoholic drinks form the basis of a massive global industry dominated by a few large firms. Each year 1.5×10^7 tonnes of malting barley are produced,[1] of which almost 94% is used in brewing and 5% in distilling.[2] Annual world production of beer is over 1.3×10^{11} litres with 36% of that coming from just ten companies.[3] The top ten spirits firms sell a total of about 2.8×10^9 litres annually and account for some 60% of the world's branded spirits.[4] The companies in the industry are marketing rather than production driven, but to a large extent owe the continued success of their brands to the technologies they have evolved.

Malting and brewing are probably the oldest of biotechnologies, having been practised for at least 6000 years. Distillation was being performed by Mesopotamian herbalists 5000 years ago and the Chinese had distilled alcohol from wine by the fourth century AD.[5] Industrialisation of the processes began in England in the 18th century and resulted from increased urbanisation and concentrated population growth providing the impetus for large-scale production with a ready market.[6] Increase in scale led to a realisation of the economic advantages to be gained from greater control of the process. From this stemmed the need for an appreciation of the underlying science of alcoholic drink production to support the empirical technologies. Much has been achieved in this quest, such that today production efficiency and product consistency are of a high order. Improved technologies have been developed with, in many cases, decreased processing time and greatly increased output per employee.

This chapter begins with a brief description of the biological and chemical changes which occur in malting, brewing and distilling, goes on to consider the current technology employed in each industry and ends with a consideration of potential targets for the application of genetic manipulation of cereals within the context of businesses dependent upon public confidence in their brands for continued success.

9.2   Fundamentals of malting, brewing and distilling

Fermentation of a sugary feedstock is at the heart of alcoholic drink production. When the drink is based on cereals, then the feedstock may originate in malted or unmalted grain. Malting is the controlled, limited germination of cereal grains (usually barley) which are then dried to give a friable, easily processed source of fermentable sugars and other yeast nutrients.

The process begins when the grain is hydrated thus triggering germination. Hormones, including gibberellic acid (GA_3) are produced by the embryo stimulating the secretion of hydrolytic enzymes (notably amylases, glucanases, pentosanases and proteases) by the scutellum and aleurone layer. These migrate to the starchy endosperm causing the progressive breakdown with time of its structural components along the barley kernel from the proximal (embryo) end to the distal end. This breakdown, termed modification, involves degradation of most of the β-glucan (primarily mixed link $\beta(1 \rightarrow 3;\ 1 \rightarrow 4)$), arabinoxylan and protein of the cell walls, together with more than half of the protein in which the starch grains are embedded inside the endosperm cells. In malting, unlike agriculture, germination is halted by drying the grain before significant (<20%) starch degradation can occur, limiting plant development to growth of the coleoptile (acrospire) beneath the husk and emergence of rootlets. Drying also causes partial or total inactivation of enzymes and the generation of colour and flavour through interactions of amino compounds and sugars in Maillard reactions.

Brewing involves the warm aqueous extraction of ground malt (often supplemented by other sources of carbohydrate) followed by separation of the soluble extract (wort), boiling with hops and fermentation by yeast to give a complex alcoholic solution which is clarified and sold as a clear, sparkling beverage with an ethanol content of ca. 3–10% v/v. The extraction process (mashing) continues the enzyme-catalysed breakdown of starch, proteins and other substances begun in malting. The starch is broken down by the combined actions of α-amylase which arises *de novo* during germination and β-amylase which is already present in the endosperm of mature barley. α-Amylase degrades amylose and amylopectin chains to short-chain dextrins and β-amylase attacks the non-reducing ends of the dextrins as these are exposed, yielding the disaccharide maltose. The primarily straight chain $\alpha(1 \rightarrow 4)$ linked amylose is almost completely degraded, whilst the extensively branched amylopectin leaves a residue of complex dextrinous material. This is because the two amylases are unable to hydrolyse $\alpha(1 \rightarrow 6)$ linkages and the debranching enzyme from malt has little activity under mashing conditions.[7] Normally about

20% of the starch sugar remains as dextrins. Only the large starch granules (20–50 μm) which make up 85% by weight of total starch in barley are readily degraded at normal mashing temperatures (ca. 65°C). Small starch granules (2–10 μm) which have survived malting are not gelatinised at such temperatures and may escape conversion. Malt amylases are unstable at mashing temperatures, particularly β-amylase which is largely destroyed by the end of mashing.

The proteases and carboxypeptidases which operate during malting continue during mashing yielding polypeptides and amino acids respectively. Brewers do not seek total breakdown of proteins. Significant proteolysis is necessary to generate amino acids for yeast nutrition and to remove the bulk of proteins which may later contribute to beer haze, but sufficient polypeptide needs to be conserved in order to allow foam formation when beer is served. The lability of malt proteases has led to the frequent practice of starting the mash at a temperature some 15°C lower than that of starch gelatinisation to allow adequate proteolysis. Degradation of any remaining β-glucan is also more efficient at this lower temperature because of the heat sensitivity of malt endo-β-glucanase. The viscosity of wort (and of beer) is to a large extent determined by the amount of β-glucan present. Too much can lead to difficulties with wort separation and beer filtration. Heat tolerant β-glucanases of microbial origin are sometimes added to the mash (or beer) if problems are encountered.[8]

Malt proanthocyanidins (polyphenols) are extracted in mashing and, together with proteins, can subsequently form hazes in beer during storage. Boiling the wort with hops after separation from the grain residue serves to precipitate most of the proanthocyanidins as complexes with coagulated protein. During boiling, the humulones, or α-acids, from the added hops are isomerised into the more bitter and soluble iso-α-acids. Boiling also serves to inactivate remaining enzymes, sterilise the wort and drive off undesirable volatiles from hops and grain. Some brewers regard dimethyl sulphide, which has the aroma of sweetcorn, in this latter category and strive to eliminate it during boiling. Others regard the compound as a critical lager component and endeavour to retain sufficient such that the level in the final beer will be above taste threshold. Dimethyl sulphide arises principally from the thermal degradation of the malt derived sulphonium compound S-methylmethionine.[9]

In fermentation of the boiled, clarified wort, sugars, amino acids and other nutrients are metabolised by brewers' yeast, *Saccharomyces cerevisiae*, to yield new yeast mass, with ethanol and carbon dioxide as major end products. Hundreds of other compounds which give flavour to the beer are also excreted by yeast. These include esters, alcohols, aldehydes, ketones, sulphur compounds, organic acids and fatty acids. Not all of these compounds are desirable in beer. The vicinal diketone 2,3-butanedione (diacetyl), which is produced by the non-enzymic decarboxylation of α-acetolactate excreted by yeast, is invariably undesirable in lagers. Its reduction below flavour threshold forms one of the prime determinants in the conduct of lager fermentation and maturation.

Beer deteriorates with time. Pasteurisation makes it microbiologically stable but haze formation, even when extensive measures are taken to remove the proanthocyanidin and protein building blocks, occurs as it ages. Flavour deterioration will also take place with the development of stale flavours due to the build-up of carbonyl compounds in the beer.

Distillation of fermented liquids produces spirits with an elevated alcohol content by a process of evaporation and condensation. The basic biochemistry of mashing and fermentation is the same in spirit production as it is in beer production. There are, however, qualitative and quantitative differences in raw materials and processing conditions which influence enzyme activity and yeast behaviour. Perhaps the most significant of these differences is that, as distillers' worts are not boiled before fermentation, enzymes from the malt continue to degrade dextrins yielding more fermentable sugars and hence, alcohol. No boiling also means that micro-organisms other than yeast are more likely to play a part in distillery fermentations than they are in brewery fermentations where every effort is taken to eliminate them. The majority of the flavour compounds in spirits are produced by yeast during fermentation[10] and are concentrated during distillation. The composition of the raw spirit will depend on the technical details of the distillation, e.g. whether it is 'pot' or 'continuous' (see Section 9.5.4), which fractions are retained and the shape and dimensions of the still. The composition of the final spirit is modified during maturation where storage in oak barrels adds colour and leads to both the release of compounds positive to flavour such as furfural and tannins, and the absorption of undesirables (principally sulphur compounds).[11]

9.3   Malting industry: current practice

Conventionally, malting is divided into three stages; steeping, germination and kilning. In current practice this division has become increasingly blurred and may now be regarded as an outmoded, if still convenient, classification. There is also rather more to commercial malt production (Fig. 9.1). Selection and pre-treatment of the grain and finishing of the kilned malt are integral parts of the process.

9.3.1   Importance of barley variety
Pre-eminent amongst cereals which are malted is barley. Not just any barley will do. Rather, defined 'malting varieties' are used. These varieties have been produced by classical breeding techniques and possess 'malting quality'. The broad criteria which need to be met for a barley sample to qualify as being of malting quality in the UK are listed in Table 9.1. These requirements characterise a barley which can be easily, rapidly and consistently converted into malt capable of yielding a high soluble extract of the required composition for beer or spirit production. Additional requirements, many of which are not

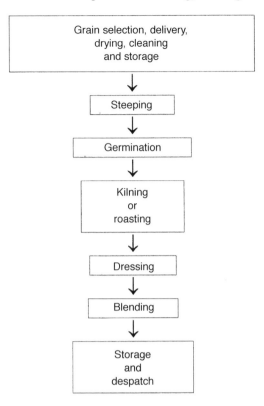

Fig. 9.1   Malt production.

currently met and may only become available through genetic modification, are discussed in Section 9.6.

Potential malting varieties produced in breeding programmes are subject to micromalting and eventually production-scale trials. A dozen barley generations may be grown and tested before a new variety can be released. Intensive breeding programmes are in place around the world, both for two-rowed barleys (which are the only type malted in Europe) and for six-rowed varieties, which predominate in the USA and Canada. Barleys which make the grade as possessing malting quality appear on national lists and it is from these lists that the maltster, brewer and distiller select varieties. Malting barleys generally yield fewer tonnes per hectare than do feed varieties. Consequently, it is common to pay a price premium to farmers for malting varieties to encourage planting. Not that once grown, a given batch will necessarily be accepted by the maltster on variety alone. It must also meet the full requirements of malting quality as listed in Table 9.1.

The turnover in varieties has become increasingly rapid in recent years as varieties possessing acceptable malting quality combined with increased yield per hectare have been introduced. Competition between breeders now leads to a

Table 9.1   Required barley characteristics for malting

Characteristic	Benefit to the maltster
Varietal purity (ca. 99% of correct single variety)	Ease and uniformity of malting
Sound grain (corns substantially undamaged, with minimal fungal contamination, absence of pre-germination, free from contaminants such as insects, stones, etc.)	Avoids problems in malting and maximises yield
Grain endosperm white and mealy rather than translucent or steely	Maximises extract and eases malting
Germinative capacity not less than 99% (virtually all grains are alive)	Maximises uniformity and extract
Germinative energy not less than 98% at time of malting (nearly all grains will germinate under controlled conditions, i.e. minimum dormancy)	Maximises uniformity and extract
Uniform plump grain (weight of 1000 corns 35–45 g depending upon variety with 90% of width 2.5 mm or greater)	Maximises extract
Appropriate nitrogen content (ca. 1.5–1.7% w/w on a dry basis)	Maximises extract
Moisture content (ca. 16%)	Indicates correct handling by farmer and is the basis of dry matter purchased
Grain capable of hydrating quickly	Accelerates malting
Grain of an appropriate structural composition and capable of developing the required enzyme complement	Necessity of malting

rapid succession of new varieties which become transiently popular and then, within the space of 4–5 years, virtually cease to be grown as they are agronomically outclassed by yet newer varieties.

9.3.2   Pre-treatment of grain

Barley is delivered to maltings in bulk, either direct from the farm or via grain merchants. It is tested before acceptance (primarily for purity, viability, nitrogen and moisture content) and passed to the undried (green) barley store. Drying of the grain to 12% w/w moisture or less is carried out as soon as practicable after delivery. Cleaning of the grain to remove impurities such as stones, earth, dust, nails as well as broken grains, husks, etc. is usually performed after drying.

Grain is invariably infested with microbes. Amongst other negative effects, these microbes (a) compete with the grain for oxygen in germination,[12] (b) produce metabolites which induce beer made from grain on which they have been deposited to spew out foam (gush) on opening the container in which the beer is held,[13] and (c) even when dead, cause haze in beer.[14] There are no practicable measures for removing these microbes from barley. The smallest corns of less than 2.2 mm width (screenings) are removed before malting. The grain is stored in bins and silos which necessarily contain a number of different batches of barley.

9.3.3   Steeping

Steeping involves increasing the moisture content of the grain to around 42–48% w/w. This is usually achieved by immersing the grain in water but, in some systems, involves spraying water over it. Before the grain is steeped, it is tested to ensure that it has matured sufficiently during storage for dormancy to have declined so that ready germination can occur. Dormancy can be a problem to the maltster when cool and wet conditions during growth lead to its extension.

Some maltsters apply an additional treatment to the grain immediately prior to steeping by subjecting it to a limited degree of battering or 'abrasion'. The mechanism of action of this treatment, which can lead to more rapid malting, is disputed[15,16] and after once gaining transient popularity (particularly amongst brewer-maltsters[17]) abrasion is now rarely performed.[18]

During steeping the grain swells by about 25% and softens as cell metabolism recommences. Water uptake is initially rapid and then gradually plateaus out. The rate of hydration is dependent upon variety, grain sample, corn size, nitrogen content, temperature and other factors.[18] As the immersion period progresses, the steep water becomes discoloured due to the presence of dissolved materials and microbes from the outer layers of the grain and is accordingly changed at least once. A temperature of around 16°C is commonly used for steeping as a compromise between the physiological optimum and the need to hydrate the grain quickly.[18] It is common practice for air to be blown through the immersed grain[19] and for 'air rests' to be incorporated into the steeping programme. This latter procedure, which involves draining off the water and leaving the grain exposed to air for a period of 8–24 hours before re-immersion, was initially advocated at the turn of the 19th century as a means of speeding up malting[20] but was little utilised at the time as such measures were regarded with suspicion. Air rests gained popularity in the 1950s with the discovery that the technique could alleviate the newly recognised phenomenon of 'water sensitivity' in which germination is inhibited by the presence of excess water.[21]

When the grain has reached the required moisture content, the water is drained off for a final time. In most malting plants the grain is then transferred (cast) to the germination vessel but in some designs, steeping and germination are carried out in the same vessel. Indeed, the demarcation between steeping and germination becomes almost notional in some plants, where the visible sign of germination (the appearance of a white 'chit' at the end of each grain) becomes evident at the end of steeping.

9.3.4   Germination

Although malting is frequently spoken of by maltsters as a natural process,[22] it hardly qualifies for that description. Successful germination to the maltster involves the controlled digestion of the endosperm with the minimal growth of new tissue. Such a transformation is only possible with human intervention.

Various additives have been used in malting, either on experimental or production scale, in order to influence grain metabolism for the benefit of the maltster. These additives have often been most effectively applied at the start of germination for such purposes as control of the microbial population, acceleration of malting, reduction in rootlet growth and control of nitrogen solubilisation. In modern practice, probably only gibberellic acid (GA_3) remains relatively widely used. This hormone is usually applied as a sprayed solution (ca. 0.1–0.25 mg GA_3/kg barley) to supplement the natural level in the grain. Within that concentration range, good quality malts may be produced in a shorter time, with smaller malting losses, as modification is enhanced to a greater relative extent than is growth.[18] The treatment is particularly useful with otherwise unpromising raw material.

Potassium bromate, when applied to germinating barley, has the effect of reducing respiration, rootlet growth and levels of soluble nitrogen. Initially used alone, with the prime objective of increasing yield, the salt found particular utility when applied in conjunction with gibberellic acid in order to mollify the excessive heat output and protein breakdown which could result from use of the hormone alone. Use of the two additives in tandem was a necessity in malting plants where temperature control was poor. This was true of all 'floor maltings' where the grain was spread thinly over a large area of a wooden floor to germinate and where temperature control was limited to adjusting the depth of grain on the floor (ca. 0.1–0.3 m) and opening and closing the small shuttered windows of the malthouse. Few floor maltings now remain and in modern 'pneumatic maltings' good temperature control is exercised by forcing humid air at the desired temperature through the bed of grain (ca. two metres deep) which is turned mechanically. Accordingly, the use of bromate in malting is now much reduced.

Ideas on the optimum temperature for germination have changed over the years. One hundred years ago, long cool germination at about 12°C for up to three weeks was the norm and malting was discontinued in the warmer months. This was at least in part due to deficiencies in plant where higher temperatures could not be risked because of the fear of over-heating and bolting. Over the years, better control of the process coupled with the wish to accelerate production have led to the use of higher temperatures (16–20°C) which, together with the use of gibberellic acid, allows germination to be complete in 3–5 days. Not that in practice germination is carried out at a single temperature. In commercial malting, temperatures are altered depending upon the type of malt required and also with the aim of regulating malt composition and adjusting for variations in raw material. Maximising both extract and homogeneity, such that each grain is modified as near as possible to the same extent, is the prime objective of the maltster.

9.3.5   Kilning

Kilning involves the drying of germinated barley in a forced flow of hot air. Choice of conditions will depend upon the type of malt being produced. For most malts, the aim is to reduce moisture to below 5% allowing storage for extended periods. It is usual to start kilning with air at a temperature of 50–60°C as it enters the grain bed. Water is then removed at an approximately linear rate until about half of the initial moisture remains. The airflow will then be reduced and the temperature raised to facilitate further drying. When the moisture has fallen to 5–8% the curing stage is reached with further reductions in airflow and increases in temperature. Curing conditions largely determine the extent of colour and flavour development and enzyme survival. With some lager malts, curing temperature may be as low as 65°C, whilst for some darker malts temperatures above 100°C are used. Some special malts are finished in roasting drums.

Modern kilns employ indirect firing in which passage of fuel combustion products through the grain bed is avoided. Furnace gases are used to heat radiators, which, in turn, heat the air for kilning. This procedure was adopted when it was found that carcinogenic nitrosamines were produced when oxides of nitrogen (NO and NO_2) in the kiln gases interacted with amines in the malt.[23]

9.3.6   Dressing, blending, storage and despatch

After kilning, the malt is dressed by a process of agitation, sieving and aspiration which removes the brittle rootlets, dust and broken corns. Freshly dressed malt is mixed with a bulk of malt which shows similar analysis. Further blending may be carried out to ensure uniformity of supply. In the UK, it is usual for blending to involve only malts made from a single variety or small group of varieties of barley whereas in North America, blending of widely different varieties, even two- and six-rowed barleys, may be regarded as acceptable. Malt is usually stored in bulk in water- and air-tight bins or silos. It is usual practice to store white malt (see Table 9.2) for at least three weeks before it is considered suitable for brewing. Malt is despatched to customers, usually in bulk, after a final cleaning and analytical check.

9.3.7   By-products

Malting produces large amounts of liquid effluent, mainly in the form of used steep water, with a high biological oxygen demand. This must be treated on-site

Table 9.2   Types of malt

Type of malt	Characteristics
Brewers' white malts: Ale and lager (pilsner)	Traditionally, European pilsner malt was purposely under-modified. Today, lager and ale malts are often equally well modified and differ only in the more developed colour and flavour of the latter. The lower kilning temperature of lager malt often leads to the survival of S-methylmethionine giving rise during brewing to dimethyl sulphide, a characteristic flavour component in many of the better lagers
Specialist white malts	So-called proteolytic, enzymic and acid malts enriched in lactic acid in order to reduce mash pH and thus approach the optima for amylolysis and proteolysis are used by some brewers as part of the grist
Brewers' coloured malts: Conventionally kilned	These include Vienna malts used in the production of middle-European golden brown lagers and the much darker and more flavoursome Munich malts used in dark lager production
Roasted	Amber, brown, chocolate and black malts in ascending order of colour are used in ale and stout production. Caramel and crystal malts are roasted wet so as to cause liquefaction, followed by crystallisation on cooling. They give luscious flavours when used at low levels in ales and lagers
Distillers' malts	Maximising spirit yield and imparting a degree of peatiness dictates the requirements of the malt whisky distillers. These objectives are achieved by using two-rowed barleys of moderate nitrogen content dried at low temperatures with peat smoke passed through the malt. Grain distillers using large quantities of unmalted adjuncts require malts with high levels of diastatic enzymes and soluble nitrogen. Hence the use of very lightly kilned malts and, in some cases, unkilned 'green' malt
Non-barley malts	Malt made from wheat, rye, triticale, oats, sorghum, maize, millet and rice have found limited use in the manufacture of beer and spirits. Wheat malt is used in the traditional wheat beers of Germany and Belgium. Sorghum malt, which needs to be germinated at high temperatures, is used for brewing in Africa

or through a local sewage works before discharge into water courses or the sea. Dried rootlets, often mixed with grain dust and pelletised, are sold for animal feed as a by-product of malting.

9.3.8 Types of malt
The main categories of malt, their characteristics and uses are summarised in Table 9.2.

9.3.9 Alternatives to malt
Unmalted cereals (adjuncts) are commonly used as part of the recipe in the manufacture of beers and spirits. The extent of usage varies widely. Malt is the only source of extract allowed in the production of malt whisky in Scotland as it is for beer manufacture in parts of Germany. In the USA the average brewers' grist (raw material mix) contains 38% unmalted adjunct whereas in the UK the figure is nearer 20%.[24] Beers have been produced from 95% raw barley and 5% malt with the help of microbial enzymes in the mash. Similarly high proportions of unmalted cereals are used in producing grain whiskies around the world.

Reasons for using unmalted cereals are varied. In the manufacture of some spirits, such as vodka and gin, the virtual absence of flavour in the distilled neutral alcohol produced from fermented unmalted cereals is a pre-requisite and use of malt is not appropriate. Bourbon, Canadian whiskey and blended Scotch whisky contain grain whisky as an essential element in the formulation which has become accepted by consumers. Nonetheless, the main reason for using unmalted cereals is that they provide a cheap source of extract. Certainly this is true with beer, although the extra plant and processing costs may eat into the apparent financial gain. There are also subsidiary reasons for the use of certain adjuncts in brewing. Thus, wheat may be used for its positive influence on beer foam and roasted cereals add distinctive flavours. The direct effect of using malt replacements on beer flavour is however only one of dilution; the inclusion of high levels of adjunct in the grist probably contributes much to the unrivalled blandness of some American beers.

Unmalted maize, rice, barley, wheat and sorghum all find routine use in brewing in different countries. Starch in wheat and barley may be gelatinised at mashing temperatures, but starches from the other cereals must be pre-cooked at higher temperatures to be gelatinised. Adjuncts are processed and refined to greater or lesser extents depending upon their nature and the stage at which they are used in the brewing process.[25] Popular forms are listed in Table 9.3. Maize and rice grits are the most commonly used adjuncts world-wide.

Table 9.3 Forms of adjunct

Adjunct form	Description
Flours	Milled grain
Grits	Dehusked and degermed pieces of endosperm
Torrefied/micronised cereals	Grain heated so that it expands and gelatinises to give a popcorn appearance. If heating is with hot air, then the cereal is described as torrefied; if the heat comes from infra-red radiation, then the cereal is said to be micronised
Flakes	Grits, steam cooked and passed through flaking rolls when hot
Syrups	Acid/enzyme hydrolysed starch

9.4   Brewing industry: current practice

Materials, processes, plant and products vary widely across the world. Nonetheless, the most usual general current practice for beer production may be summarised as in Fig. 9.2.

9.4.1   Raw materials

Malts and unmalted cereals available for use in brewing have been surveyed in Sections 9.3.8 and 9.3.9 respectively. Brewers have exacting (and sometimes contradictory) specifications for the malts they use, ranging from the basic requirements concerning the amount of extract they can expect to the level of individual compounds as in the case of S-methylmethionine. With unmalted cereals, extract and fermentability are usually the key parameters. Standard or recommended methods of analysis for brewing materials have been codified easing commercial transactions.[26] The most common characteristics of malt specified by brewers are listed in Table 9.4.

9.4.2   Milling

White malt, together with any coloured malts, torrefied/micronised or roast cereals used in the recipe for a particular beer (see Tables 9.2 and 9.3) is milled prior to extraction with water. The intensity of milling required is determined by the degree of modification of the malt and the plant which will subsequently be used to effect wort separation. Particle size is the important factor. Properly milled malt should yield a grist composed predominantly of finely powdered endosperm which retains more sizeable fragments of husk. The more highly modified the malt, the less intensive need be the milling to give an appropriate

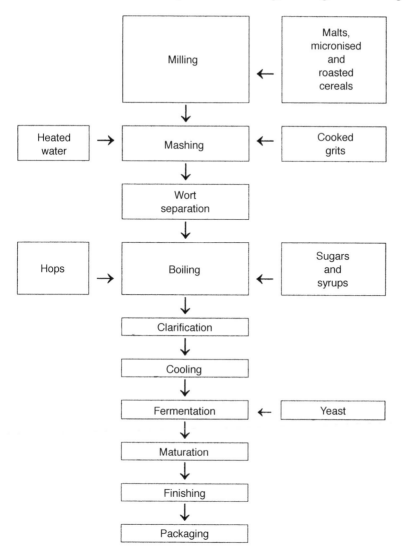

Fig. 9.2 Beer production.

particle size. Similarly, different degrees of survival of husk will be required dependent upon the extent to which this component is required to act as a filter bed. In some breweries, the malt is sprayed with a fine stream of water immediately before it enters the mill. This makes the husk less brittle and gives bigger fragments. In extreme cases, this principle is carried further and the malt is steeped in warm water and the grain 'wet milled'. Roller mills are the most usual as their crushing action suits the particle size required in the lauter tuns employed in most breweries. Hammer mills are used only with mash filters (see Section 9.4.4).

Table 9.4 Common malt characteristics specified by brewers

Malt characteristic specified	Benefit to the brewer	Barley characteristic required
High hot water extract	Maximises potential yield	Grain of malting quality with mealy endosperm and capable of generating the required enzyme complement
High friability	Maximises actual yield	Grain which can be degraded in malting and mashing to the maximum extent
High homogeneity	Maximises potential yield and ease of achieving it	Uniform grain structure and response in malting
Low β-glucan	Minimises processing problems	Endosperm cell wall easily degraded
Sufficient α-amylase and β-glucanase	Essential in mashing and further indication of appropriate malting and kilning	Grain capable of generating required enzymes
Required wort fermentability	Will yield required alcohol content	Grain yielding malt of appropriate structural and enzyme composition
Optimal level of total soluble nitrogen and free amino nitrogen	Gives required yeast growth	Appropriate grain total nitrogen
Appropriate moisture	Basis of dry matter purchase and indication of appropriate kilning	–
Appropriate colour	Defines beer colour	Appropriate grain total nitrogen
Appropriate level of S-methylmethionine	Gives required flavour to beer	Methionine methylation system operational to required degree

9.4.3 Mashing

The primary aim of mashing is to 'convert' the mash, i.e. through starch hydrolysis, to render ca. 80% of the grist soluble. There are a number of variations on the theme of mashing with regard to timing, temperatures and plant used but one system, temperature-programmed infusion mashing, is now the most common and will be described here.

After brief storage in a hopper (grist case), the milled grist is mixed with warm water (liquor) on transfer to the agitated mashing vessel (mash tun or mash mixer). Best practice is to avoid uptake of air when mashing as this minimises any contribution to later beer staling through the action of malt lipoxygenase.[27] Thus, the grist/liquor mix is introduced close to the base of the mash tun and agitation is kept to a minimum. Any cereal grits (see Table 9.3) are introduced into the mash tun after prior gelatinisation of the starch in a 'cereal cooker' at 80–100°C in the presence of either malt or a microbial source of α-amylase to give liquefaction. Other additions to the mash tun may include heat-resistant microbial enzymes to aid conversion or salts/acids to adjust mash pH towards the desired figure of around 5.5.

Mashing programmes are designed to best accommodate the properties of the enzymes provided by the malt. Thus, mashing will typically commence at a relatively low temperature (45–55°C) allowing some activity by the more heat-sensitive enzymes, e.g. protease and β-glucanase. After a stand of perhaps 30 minutes, the temperature is raised (ca. 1°C/minute) to 63–7°C to gelatinise starch and accelerate α-amylase activity. A stand of around one hour is then followed by increasing the temperature to 76°C thus halting most enzyme activity and reducing viscosity.

9.4.4   Wort separation

Traditional practice is to effect separation by carefully drawing the wort through the bed formed by the insoluble residue of the grains on the slotted plates which form the false bottom of the mash tun. This system is still used for some British ales but current practice in most breweries is to use a separate lauter tun for all types of beer. A lauter tun is a straight-sided, round vessel, equipped with rakes or knives hanging vertically from (usually) three horizontal central arms. The arms can be made to revolve around a central spindle above the slotted base of the lauter tun. The mash is transferred to the bottom of this vessel to a depth of ca. 0.5 m. The rakes are used to level the bed and to cut through it as required to ease the passage of the wort.

Operation of the lauter tun is carefully controlled in order to balance the requirements of wort clarity, maximum recovery of extract, required wort strength and maximum rate of run-off. Clarity is ensured by recycling the wort, i.e. returning the outflow to the top of the bed until it runs 'bright'. Extract is maximised by sprinkling (sparging) the mash with hot liquor at 77°C. The required strength is achieved by limiting the quantity of sparge used. Run-off rate is enhanced by the shallow bed depth, large surface area, reduced viscosity due to efficient β-glucan degradation in malting and mashing, and optimised particle size of the grain bed.

In skilled hands, using good materials, a well-designed lauter tun will typically allow efficient wort separation in about 90 minutes. The same result can be achieved in about half that time using a plate and frame mash filter in which the mash is passed through a series of polypropylene cloths under pressure. Mash filters have been around for over one hundred years but have gained popularity only in the last decade or so with the introduction of improved designs offering real economies. They are less exacting to operate than lauter tuns, particle size in particular ceasing to be important, although wort viscosity must still be minimised by ensuring low β-glucan levels.

9.4.5   Boiling

After being separated from the grains, the wort is passed to a vessel known either as the kettle or the copper, after its traditional material of construction (although stainless steel is now more usual). In the copper, the wort is boiled with hops or hop preparations for 60–90 minutes. Modern practice is to heat the copper by passing the wort through an external heat exchanger called a calandria. This gives good agitation to the wort, increasing the effectiveness of the boil and enhancing precipitation of the coagulated protein, etc., known as 'hot break' or trub. Any brewing sugars or syrups are blended into the wort in the copper.

9.4.6   Clarification and cooling

Current practice in most breweries is to remove trub, together with any residual hop material, from the wort in a device known as a whirlpool. Hot wort is pumped tangentially into a cylindrical vessel such that it swirls around causing the trub to accumulate in the middle of the vessel by centripetal force, much as spent tea leaves gather in a rotated cup. The clarified wort is drawn off from the outer rim of the base of the whirlpool.

Wort from the whirlpool is cooled to the temperature required for fermentation by passing it through a plate heat exchanger. Cooling results in further precipitation of insoluble protein and lipid material known as 'cold break'. A certain level of cold break held in suspension in the wort is beneficial to fermentation but too much will lead to subsequent beer-processing and quality problems. It is the practice in some breweries to restrict the level of suspended cold break by adding carrageenans (charged polysaccharides extracted from seaweed and known as 'copper finings') to the wort a few minutes before the end of boiling to enhance the rate of sedimentation. The effectiveness of copper finings is enhanced if wort pH is adjusted with sodium bicarbonate before addition.[28]

9.4.7   Fermentation and maturation

The cooled wort is aerated or oxygenated (usually ca. 8–16 mg/litre O_2) and inoculated (pitched) with a selected single strain of brewers' yeast (ca. 10–15 million cells/ml) *en route* to the fermenter. During fermentation, the temperature is actively controlled. Lager fermentations are carried out at lower temperatures (ca. 8–14°C) than ale fermentations (ca. 15–22°C). Yeast cell numbers increase three- to fourfold in fermentation. As the fermentable carbohydrates near exhaustion, the yeast separates either to the top or the bottom of the vessel bringing to an end 'primary fermentation'. Subsequent treatments of the green beer depend upon the type of beer and the philosophy of the brewery.

In traditional lagering practice, the fermenting wort is chilled to around 0°C whilst some yeast remains in suspension and a little fermentable carbohydrate is still available. The beer is then held for weeks or months to allow flavour maturation and precipitation of haze-forming materials which affect shelf life. Flavour maturation involves so-called 'secondary fermentation' in which the residual yeast gradually removes the undesirable flavour by-products produced earlier in the fermentation (principally diacetyl) and excretes compounds which enhance fullness (body). Secondary fermentation in a closed vessel leads to the build up of carbon dioxide in the beer to give it its characteristic sparkle. In many modern lager breweries there is no secondary fermentation period. Instead primary fermentation is continued until carbohydrate is exhausted and diacetyl levels have fallen to acceptable levels (end-fermentation). Maturation is then reduced to a period of perhaps one week's cold storage (less than 0°C) after yeast removal (often assisted by centrifugation) to allow precipitation of haze-forming materials.

At the end of primary fermentation, traditional 'cask conditioned' ales undergo 'warm conditioning' in the presence of yeast at ca. 13°C in their final package, often with sugar addition (primings) to encourage secondary fermentation. 'Isinglass finings' (processed collagen from the swim bladders of certain tropical fish) and 'auxiliary finings' (negatively charged polysaccharides) are added to the cask to encourage yeast and protein precipitation respectively in these beers. 'Brewery conditioned' or 'keg' ales are 'cold conditioned' (less than 0°C) in the absence of yeast for periods as short as three days.

In traditional processes, fermentation and maturation are carried out in separate vessels. In modern practice this may also be the case, but in some breweries the stages are combined in one temperature-controlled vessel. Fermentation vessels come in various designs, but current practice in most breweries is to install vertically mounted cylindroconical vessels from which yeast may be conveniently removed from the cone at the base. The use of these vessels, sometimes in conjunction with silicone antifoams, has blurred the distinction between the characteristics of top yeasts which rise to the surface and bottom yeasts which sink, for ale and lager brewing respectively. Virtually all beer is made by batch processes. Continuous fermentation and maturation, although much researched, are little practised.

9.4.8   Finishing and packaging

Many beers depend not only upon cold storage for precipitation of haze-forming materials but also have additions made to them in order further to enhance shelf life. These treatments include silica gels and gallotannins to remove polypeptides and polyvinyl pyrrolidine (PVPP) to remove tannins. The additions may be made during maturation or prior to filtration, or may indeed form part of the filter medium. The addition of papain to improve beer shelf life, though once popular, is now out of favour because of its negative effect on beer foam.

Following maturation/cold conditioning, beer is generally filtered to remove remaining suspended material. Most commonly, the beer is mixed either with kieselguhr (a soft powder consisting of the fossil remains of unicellular algae) or perlite (vitreous volcanic ash) which help to trap particles and keep the filter bed from clogging, and then passed through a plate and frame filter at 0°C. The 'bright beer' is collected in a holding tank and the carbon dioxide level is adjusted ready for packaging. Great precautions are taken to ensure that the oxygen level in beer is maintained as low as possible (less than 0.1 mg/litre) during filtration and packaging in order to prevent oxidation of the beer. To this end some brewers add sulphur dioxide or ascorbic acid to beer at this stage to act as antioxidants. Other materials which may be added prior to filtration include caramel to adjust colour, isomerised hop extract to meet bitterness specifications and propylene glycol alginate to help preserve beer foam. Foam may also be enhanced dramatically by dissolving gaseous nitrogen in the beer.

Traditionally, brewers talk of the strength of beer in terms of 'original gravity', meaning the specific gravity of the wort before fermentation. Original gravity is determined primarily by the concentration of dissolved carbohydrate in the wort. Thus the higher the gravity, the more alcohol that will be produced in fermentation and the stronger the beer. Until some 25 years ago, beer was nearly always produced at the strength at which it was to be consumed. Many beers are now produced at 'high gravity', up to twice as strong as 'sales gravity'. These beers are diluted with deaerated, carbonated water and adjusted for colour and bitterness by suitable additions just prior to packaging.

9.4.9   By-products
The residue after wort separation, known as 'spent grains', is sold without additional treatment as cattle food. Carbon dioxide may also be collected and used in the brewery or sold. Spent yeast is sold for making yeast extract.

9.4.10   Beer types
Despite the convergence of brewing techniques internationally, which has led to a blurring in historical distinctions between the methods of manufacture of beers, a number of styles may still be identified (Table 9.5).

9.5   Distilling industry: current practice

Distilled spirits derived from cereals may be classified into three groups.

1.   Those based on essentially unflavoured neutral alcohol, e.g. vodka and the rice wine derived shochu.

Table 9.5 Types of beer

Types of beer	Characteristics
Ales	Typically 3–5% alcohol. Mild ale is sweet and dark, pale ale is bitter, light brown and stronger. Cask ales are low in carbon dioxide (ca. 2 g/litre) and have a short (4 weeks) shelf life. Keg and bottled ales are matured and stabilised in bulk at low temperatures and have carbon dioxide (ca. 4–6 g/litre) and shelf life specifications similar to lager (26–52 weeks). Specialist ales include barley wine with up to 10% alcohol. Widget and nitrokeg ales are low carbon dioxide, chilled and filtered beers nitrogenated to give pressure in the container with the bonus of a creamy head. They mimic traditional cask beer but should not be mistaken for the real thing.
Lagers	Typically 3.5–5.5% alcohol. Many straw-coloured beers from around the world fall into this category, some very light in flavour, others full-bodied and satisfying. Traditional malty dark lagers may still be found in Germany and Belgium.
Stouts	May either be sweet (3–4% alcohol) or bitter (4–7% alcohol) depending upon the amount of free sugar they contain and the quantity of hops used in their brewing. They are black in colour from the roast barley or dark malt component of the grist.
Diet or 'lite' beers	These beers (ales and lagers) are characterised by a fuller conversion of dextrins to free sugar which can then be fermented to alcohol. Thus, they have a lower carbohydrate content than normal beers. Production involves mashing to give maximum fermentability and the addition of debranching enzymes to the fermenter.
Wheat beers	May be cloudy (due to presence of yeast) or bright. The Bavarian version has a phenolic taste due to excretion of 2-vinyl guaiacol by the yeast used. Belgian versions are clean and fresh tasting.
Alcohol-free and low-alcohol beers	Definition of these depends upon legislation in individual countries. Typically alcohol-free means <0.05%, low-alcohol <1.2%. They are produced either by restricted fermentation or by alcohol removal from standard beers.
Ice beer/dry beers, etc.	Marketing spin on straw-coloured lager.

2. Products into which some secondary flavouring material has been introduced into the neutral alcohol either during or immediately after distillation, e.g. gin, schnapps and many liqueurs. These beverages generally require no ageing.
3. Drinks where much of the characteristic flavour is derived from the cereal raw material and/or from the ageing process, e.g. whisky and kornbranntwein.

Beverages based on neutral spirits are relatively simple products often made from the cheapest source of starch available in a given country. Only whisky, the most complex cereal-based spirit, the production of which exemplifies in its fullest form the operations involved in distilled beverage manufacture, will be considered in any detail here. A number of distinct whiskies exist, determined primarily by the cereal or cereals used and the details of the production process. There are five major whisky-producing countries in the world: Scotland, Ireland, Canada, USA and Japan. Four of these countries have evolved traditional products which have their origin in local agriculture. The Japanese whisky industry has been created by entrepreneurship in the last 75 years or so. It should be noted that the spelling whisky is used throughout the English-speaking world with the exception of Ireland, the USA and for some Canadian products where the variant whiskey is used. A generalised scheme for whisky production is shown in Fig. 9.3.

9.5.1   Raw materials

The essential characteristics of distillers' malt have been described in Section 9.3.8 and the forms which unmalted cereals can take in Section 9.3.9. Whilst malted barley is the predominant ingredient for beer, other cereals come into prominence with whisky. The identity of cereals used in whisky production, the proportions in which they are used and whether or not they need to be malted, are often defined in legislation and/or through tradition and practice.

Scotch whisky may be produced only in Scotland. Scotch malt whisky is made from lightly kilned, peated malted barley alone. Malt whiskies are also produced in Japan.[29] Scotch grain whisky is produced from cooked, unmalted maize or wheat (80–95%) and malted barley (5–20%). Grain whisky plants operate on a much bigger scale than malt whisky distilleries producing about ten times the volume of spirit. European Union preferences have made it more economical for Scotch grain whisky distillers to use wheat rather than maize which had long been the cereal of choice.[30] Almost all the grain whisky produced in Scotland is mixed with malt whisky and sold as a proprietary blend. Golden Promise was for a long period the preferred barley variety for Scotch malt whisky production but it has now been supplanted by newer varieties which give increased spirit yields. Spirit yield is dependent upon the amount and fermentability of the extract available from the malt and the quantity of extract lost in new yeast mass. Whisky distillers require barley varieties which contain low levels of cyanogenic glucosides. These compounds are converted during mashing and fermentation into hydrogen cyanide which, during distillation in contact with copper surfaces, yield ethyl carbamate (urethane), a carcinogen. The use of exogenous gibberellic acid is not permitted in the manufacture of malt for use in Scotch whisky.

By definition, American bourbon whiskey contains at least 51% maize and often up to 80%.[31] Rye, six-row barley malt containing high levels of starch degrading enzymes and occasionally wheat are also included in the grist. Rye

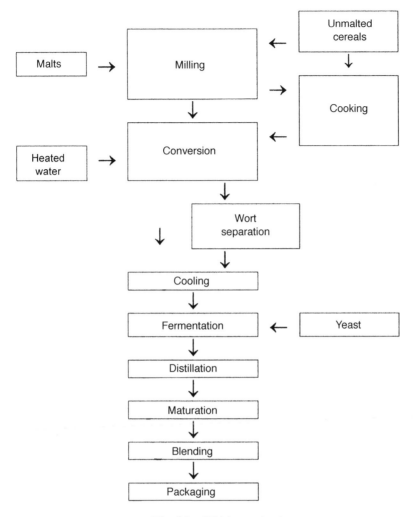

Fig. 9.3 Whisky production.

whiskey grist contains a minimum of 51% unmalted rye together with wheat and rye malts. Irish whiskey is made from a grist containing 40–60% raw barley with most of the balance made up of unpeated barley malt with a small amount of wheat or oats. Distillers of Canadian whiskies are not regulated as to the identity or proportions of cereals they may use. In practice, however, these products are usually based on maize with heavier 'flavouring whiskies' from rye, barley malt and rye malt blended in at about 10% to build up complexity.[32]

9.5.2   Milling, cooking, conversion and separation

Methods of grinding the starchy raw materials to a flour and converting them into fermentable extract are determined by the nature of the material used. Four-roll mills may be used with friable malts, whereas hammer milling is common for unmalted cereals.

Repeated infusion mashing at four different temperatures (e.g. 63°C, 70°C, 80°C and 90°C) over a total period of 6–8 hours, with the final water used for mashing-in the next batch is usual in Scotch malt whisky production. Although traditional deep mash tuns with rotatory agitators are still used, lauter tuns equipped with knives and sparge rings have been introduced in some distilleries.[33]

Where unmalted cereals are used, it is usual to gelatinise the material by cooking before conversion to fermentable sugars in the mash. The only exception to this procedure is with Irish whiskey where finely ground barley is not pre-cooked before mashing with malt. Maize or other grain is often processed in batch cookers at 120°C or higher temperatures; alternatively, continuous cooking in tubes may be employed. In some distilleries the grain is milled prior to cooking, in others the grain is cooked whole. Sour mash bourbon distilleries mix the diluted, partially clarified acid residue (setback) from the still with the maize at the start of cooking.

Conversion may be with either enzyme-rich malt or with microbial enzymes. Barley malt is the only source of enzymes allowed in Scotch grain whisky production, most Canadian distilleries use only microbial enzymes and bourbon distilleries may use either.[31,32,34] In order to maximise yield, the grist for Scotch whisky may be extracted several times, with weak worts from later extractions being used in subsequent mashes or in cereal cooking.[34]

In contrast to beer production, whisky worts are never boiled. Instead, after cooking and mashing, the cooled wort may be passed directly to the fermenters (washbacks) without first separating the grist solids (in-grains or whole mash). Alternatively, the wort is partially clarified by collection through the slotted plates of the mash or lauter tun before cooling *en route* to the fermenters. The latter process is practised with Scotch malt whisky, the former with American bourbon. The solids may or may not be separated prior to fermentation with Scotch grain whisky, depending upon the individual distillery.[30]

9.5.3   Fermentation

In Scotch whisky production a single strain distillers' yeast or a mixture of this yeast with 30–50% of yeast from a brewery fermentation is used. Distillers' yeast gives particularly good ethanol yield and carbohydrate utilisation for reasons which are poorly documented.[35] Yeast is inoculated at ca. 2×10^7 cells/ml and there is an approximately tenfold increase in the number of cells in 12 hours as the fermentation proceeds rapidly and vigorously, reaching temperatures over 30°C.[10] There is no temperature control or mixing, nor are any additions made to the fermenter. Fermentation is complete in 40–48 hours with alcohol content at ca. 6–10% (v/v). Traditionally, Scotch whisky fermentations are carried out in wooden fermenters made of larch. Many wooden vessels are still in use[36] but stainless steel fermenters are now found in many distilleries.

Large stainless steel fermenters (ca. 4000 hl) are common in bourbon distilleries although smaller open-topped fermenters are still used by some. Bourbon fermentations usually start warm, ca. 30°C, and are allowed to rise to 33–35°C. Yeast, inoculated at ca. 8×10^6 cells/ml, may come from an in-house supply or a commercial dried yeast may be used. Ethanol levels reach 8–12% (v/v).

Inoculum concentration with Canadian maize whiskey fermentations is usually much lower than that used for Scotch or bourbon fermentations at around 4×10^6 cells/ml. Fermentation lasts 72–96 hours at a controlled temperature of 26–30°C producing 10–11% v/v alcohol. Rye-based Canadian flavouring whiskey fermentations reach ca. 7% v/v alcohol. In-house yeast strains are commonly used for producing these flavouring whiskies, whereas commercially available yeast is often used for the base maize whiskey component.

9.5.4   Distillation

The contents of the fermenter are transferred to the still without the removal of yeast. There are two basic types of distillation, batch pot distillation and continuous distillation.

A pot still resembles a large onion/pear-shaped kettle with a rising swan neck. Traditionally, these stills are made of copper although stainless steel is now also being used. Heating of the still may be either by direct firing or by means of internal steam heat exchangers. Direct heating of the still yields pyrolysis products which add flavour. In distillation, vapours pass up through the swan neck and then via a connecting pipe (lyne arm) to a condenser. Still shape, which to a large extent regulates the composition of the spirit, is of major importance in determining the flavour of the whisky from a particular distillery.

Spirit from a pot still will go through at least two distillations. In Scotch malt whisky production, the first distillation (in the wash still) yields a product containing 20–30% v/v alcohol (low wines) which is distilled (in the spirit still) to give raw whisky (British Plain Spirits) of 65–70% v/v alcohol. In the production of Irish whiskey there are always three distillations giving a raw whiskey with a higher alcohol content than is the case in Scotland. The first and last fractions of the distillate (foreshots and feints respectively) are in all cases collected separately and returned for redistillation. In addition to being used for the production of malt whiskies in Scotland, Ireland and Japan, pot distillation may also form part of the process with other whiskies, e.g. for the flavouring whiskey component in Canadian whiskey. However, continuous distillation is used for the bulk volume of whisky made world-wide.

Continuous stills originate from the ingenious patent still invented by Aeneas Coffey in the early 19th century. This still consists of two columns called the analyser and the rectifier. Preheated fermented liquid (wash) is introduced into the top of the analyser with steam entering at the bottom. The steam strips out volatiles from the wash and the vapour is fed to the bottom of the rectifier which contains the pipe carrying the incoming cold wash. The vapour rises up the rectifier and condenses, thus heating the wash which is, in turn, on its way to the top of the analyser. Spirit is drawn from the rectifier at a point selected to give the required composition. Products containing up to 95% alcohol with only low levels of flavour compounds (congeners) are obtained in this way.

A refinement of the continuous patent still, used in distilleries where a spirit of particularly neutral flavour is required, introduces an intermediate column between the analyser and the rectifier. Condensed spirit from the analyser is fed to the top of the intermediate (extraction) column and is diluted to about 15% v/v alcohol with water. Because of their relative insolubility, unwanted congeners can be stripped from the liquid by steam and collected from the top of the column, while purified ethanol is withdrawn from the bottom and fed to the rectifier where it is concentrated. The operation is known as extractive distillation.

9.5.5 Maturation, blending and packaging

The minimum maturation period (sometimes enforced by legislation) for most whiskies is three years but maturation can be continued for over 30 years. Traditionally, Scotch whisky was matured in sherry butts and still is with some malt brands, but bourbon barrels are most often used. Bourbon is matured in new barrels. The inner surface of the barrels is often charred before use enhancing absorption of sulphur compounds from the whiskey.

During maturation the fiery, colourless spirit darkens and mellows. Almost all of the whiskies which are sold are blends in that they are mixtures of whiskies from more than one production run. This applies even to 'single malt whiskies' which are the product of a single malt whisky distillery. They are not to be confused with 'blended Scotch whiskies' where 30–40 different mature malt whiskies from various distilleries will be blended with two or three grain whiskies. In Canada a similar post-maturation blending approach is also used although it is also frequent practice to select raw whiskies prior to maturation, mix them and mature them together.[32] A further blending of the mixed whiskies is then carried out when they have matured.

Different degrees of latitude are shown in what additions may be made to matured whisky. Caramel to adjust colour is the only additive allowed with Scotch whisky and even that is not allowed in bourbon. Canadian distillers on the other hand, are allowed to add small amounts of flavourings or 'blenders' including wine, sherry, brandy, rum, bourbon and malt whisky as well as caramel. Blended whiskies are usually stored for several months to 'marry'. The final operation is to chill the whisky down to less than 10°C, filter to remove precipitated material and bottle the diluted spirit at sales strength, usually 40% v/v alcohol.

9.5.6 Neutral and compounded spirits

Grain spirits other than whisky are based on neutral alcohol manufactured in a continuous still, often by the process of extractive distillation described above.

The cheapest of raw materials are often used as are heat-stable microbial enzymes for conversion. Vodka is the spirit of most commercial importance. It may be derived from a variety of raw materials including potatoes but for Western markets it is usually distilled from grain (frequently rye). The high strength distillate is charcoal filtered, diluted with water, has a little glycerine added for body and is bottled. Gin is also produced from neutral spirit which is often grain derived. It receives its distinctive flavour by being redistilled with a number of flavouring materials (botanicals), most importantly juniper. Alternatively, gin may be produced by 'cold compounding' where the neutral spirit is blended together with a concentrated distillate of botanicals. As with vodka, there is no maturation period prior to dilution and bottling. Schnapps may be distilled from grain or potatoes and is usually flavoured with caraway or aniseed. Akvavit or aquavit is its Scandinavian name.

9.5.7   By-products
As with breweries, wort separation (if practised) in distilling gives a residue (draff) and this, together with residues from both pot (pot ale) and continuous (spent wash) distillations is sold, after drying, under the name 'dark grains' to animal feed companies. Carbon dioxide is collected, scrubbed, compressed and liquefied. Fusel oil from the rectifying column in continuous stills may be sold for use in perfumery.[37]

9.6   Summary: limitations in current practice and the role of biotechnology

All the indications are that barley will remain as the major cereal for brewing in developed countries. New varieties of barley will need to continue to combine good agronomic characteristics with malting quality. The popularity of individual unmalted cereals used in brewing and distilling will continue to be dependent upon the market price for these commodities.

As a broad general requirement, all three industries continue to seek ever lower processing costs without affecting quality as perceived by the consumer. Cost savings in terms of revenue and capital expenditure and through reduced waste are all sought in amongst other ways by increasing the efficiency of raw material usage. Because the industries are so closely interlinked, many problems are common to all three. Problems related to cereal quality and characteristics which have been identified[18,38,39] may be listed as follows:

- With current barley breeding programmes, malting quality and suitability for brewing and distilling can only be assessed relatively late in the breeding cycle.
- Dormancy and water sensitivity of barley continues to be a problem in some years and with some varieties.

- Drying barley to prevent deterioration in storage is energy intensive and expensive.
- The balance of structural components and enzyme levels in germinating barley and malt is still far from optimal. Maltsters require to achieve quicker and more even modification. Brewers seek more easily processed raw material yielding more soluble extract in mashing and giving beers with longer shelf lives and better foam characteristics. Malt distillers require continued improvement in spirit yield. Grain distillers seek maximum enzyme levels.
- Half the barley biomass is lost during malting or discarded as waste in beer and spirit production. This needs to be reduced.
- Malt analysis provides only a very imperfect guide to brewhouse performance and beer and spirit quality.
- For economic reasons, it is desirable to use indigenous sorghum for beer production in Africa. Sorghum is not, however, a good cereal for producing clear, European-style beers. Malting of sorghum incurs high losses of material. Sorghum malt contains low enzyme levels and starch with a high gelatinisation temperature.

These problems/needs should be seen in the context of the environment in which the major companies in the drinks industry operate. Globalisation of the brewing and distilling industries with the development of ever bigger multinational companies presents increasing problems for consistency of raw material supply if brand credibility is to be maintained internationally. The logic of company mergers also requires production facilities of ever greater batch size in order to bring the promised economies, thus putting greater emphasis on raw material consistency.

Safety and wholesomeness of food and drink continue to receive increasing attention, both through legislation and consumer interest. Any potential savings need to be offset against negative perceptions if they are not to be counter productive.

The drinks industry cultivates a traditional image, where persuading consumers of the authenticity and integrity of a company's brands is seen as of prime importance. Even when production is carried out in what are effectively modern factories, companies would prefer to be seen as craft- rather than technology-based when trying to generate positive identification with the product and appeal to the aspirant lifestyle of consumers.[40] Behind this marketing facade, major companies have in fact demonstrated a readiness to adopt new technology when a clear commercial advantage is evident but are, understandably, extremely wary of anything which may endanger the expensively generated images of their brands.

Public attitude to genetic manipulation is now seen by many in the drinks industry as just such a danger making the technology more of a threat than an opportunity. Thus, we have the following recent statement from a senior European brewing scientist.[41] 'The arrival on the market of genetically modified cereals represents a real worry for the brewer concerned to maintain the image of beer as a healthy and natural product.' Following the introduction of genetically modified maize, some brewers have sought assurances from their suppliers that nothing sent to them will contain transgenic material[42-44] and techniques for detecting admixture of Bt maize have been developed.[45,46] Perhaps surprisingly in view of the commodity status of maize, a guarantee of purity has been obtained in at least one case.[43,47] A similar lack of enthusiasm for the use of genetically modified materials in Scotch whisky production has been indicated.[48]

Despite these negative signals, targets for genetically modified barley have been identified by scientists in the industry.[38,39,49-52] Published targets together with some additional ideas inferred from other work[7,12,14,27] are summarised in Table 9.6. Progress has already been made towards achieving some of the goals,[53-57] and indeed, in the case of proanthocyanidin-free barley, a malting and agronomically acceptable variety has been produced (albeit with a gestation period of 20 years from first detection) by a classical recombination breeding programme based on chemically induced mutations.[58]

Some brewing scientists hold to the view that the industry will eventually embrace gene technology.[24] Others go further and confidently predict that a biotechnology-led revolution is just around the corner for the drinks industry.[39] Whilst long-term trends are difficult to predict, the reality is that developing consumer reaction[59,60] makes it increasingly unlikely that the drinks industry in Europe will voluntarily become associated with genetically modifed materials in the current climate. Even in the US, where consumer attitudes are at present different[61] and where genetically modified maize presumably already constitutes a large proportion of the grist in many beers and whiskies, a major company is unlikely to risk handing a potential advantage to competitors in the shape of negative publicity. They will not have forgotten the acrimonious public slanging match of claims and counter claims over the use of additives which broke out between two large US brewers only a few years ago. Nobody will want a repeat of that episode.

It is a sobering fact that, 20 years after work on the genetic modification of brewing yeast began with much initial enthusiasm,[62] the production of many strains with useful attributes[63] and the clearance of legislative hurdles,[64] no brewer anywhere in the world is using such a strain. Given the heavy downside, it is hard to see how genetically modified cereals will escape a similar fate. Indeed, now that the predicted[65] downturn in research effort in breweries generally has come to pass with increasing focus on short-term activities,[52] the task of implementing advances in biotechnology will be that much harder.

9.7   Sources of further information and advice

D.E. Briggs's books *Barley*[66] and *Malts and Malting*[18] are the two most authoritative sources on the brewers' primary raw material. Wider coverage is

Table 9.6 Targets for genetic modification of cereals for malting, brewing and distilling

Target	Benefit
Improved disease and pest resistance	Reduces chemical residues in the grain (9.3.1)
Safe barley storage without drying	Energy savings (9.3.2)
Inhibitory to grain microflora	Improved malting and brewing properties and reduced tendency for gushing (9.3.2)
Reduced husk content	Higher yields (9.3)
Control of dormancy	Improved efficiency of malting operations (9.3.3)
Quicker and more even water uptake	Faster malting (9.3.3)
Increased natural antioxidants, e.g. catechin and ferulic acid in the grain	Improved beer flavour stability (9.2, 9.4.3, 9.4.8)
Reduced levels of β-glucan in barley	Faster malting and easier wort separation and beer filtration (9.2)
Thermostable β-glucanase	More effective β-glucan breakdown in mashing (9.2, 9.4.3)
Thermostable α-amylase, β-amylase and limit dextrinase	Improved extract and fermentability (9.2, 9.4.3, 9.5.2, 9.5.3)
Modified barley endosperm structure, e.g. changed amylose/amylopectin ratio or increased ratio of large to small starch granules	Starch more easily convertible giving increased yields (9.2, 9.4.3, 9.5.2)
Increased levels of hydrophobic proteins	Improved beer foam (9.2, 9.4.8)
Reduced levels of heat-stable proteins that limit dextrinase	Improved fermentability (9.2, 9.4.3, 9.5.2, 9.5.3)
Proanthocyanidin-free barley	Better haze stability (9.2, 9.4.8, 9.6)
Blockage of methylation of methionine in germination	Lower dimethyl sulphide in lager (9.2)
Stimulation of methylation of methionine in germination	Higher dimethyl sulphide in lager (9.2)
Lower levels of malt lipoxygenase	Better control of beer staling (9.4.3)
Sorghum more amenable to malting and brewing	Improved economy of brewing in Africa (9.6.1)
Cyanogenic glycoside-free grain	Elimination of carcinogenic ethyl carbamate production in distillation (9.5.1)

Note: Numbers in parentheses indicate where in this chapter the benefits are placed in context.

given by *Cereal Science and Technology* edited by G.H. Palmer.[67] *Malting and Brewing Science*[68] by J.S. Hough, D.E. Briggs, R. Stevens and T.W. Young is the most comprehensive textbook on these subjects although a new edition is overdue. More accessible is *Brewing*[69] by T.W. Young and M.J. Lewis. A North American perspective is provided in *Handbook of Brewing*[70] edited by W.A. Hardwick, and European lager technology is given in-depth treatment in W. Kunze's *Technology: Brewing and Malting.*[71] A very readable introduction to malting and brewing is *Beer*[24] by C.W. Bamforth. There are fewer worthwhile source books on distilling. *The Science and Technologies of Whiskies*[72] edited by J.R. Piggott, R. Sharpe and R.E. Duncan gives good coverage. G.N. Bathgate's chapter in G.H. Palmer's book on cereal science[67] noted above is also worth consulting. The Proceedings of the Institute of Brewing Aviemore Conferences on Malting, Brewing and Distilling, held every four years since 1982, go some way towards filling the gap in the literature concerned with the science and technology of distilling.

Notable bodies on the technical side of the brewing industry include The European Brewery Convention (EBC), P.O. Box 510, NL-2380 BB, Zouterwoude, The Netherlands which aims to co-ordinate work carried out in different countries. The EBC holds an international congress every two years and is in the process of publishing a series of detailed *Manuals of Good Practice* on all aspects of malting and brewing. Many countries have members' organisations for brewing scientists and technologists. The Institute of Brewing, 33 Clarges Street, London, W1Y 8EE is a well-established organisation which issues a number of publications including two journals. The American Society of Brewing Chemists, 3340 Pilot Knob Road, St. Paul, MN 55121-2097, USA and The Master Brewers Association of the Americas, 2421 North Mayfair Road, Suite 310, Wauwatosa, WI 53226, USA also hold conventions and publish their proceedings.

There are numerous institutes and universities who carry out research associated with the drinks industry. Perhaps the most active are the International Centre for Brewing and Distilling, Heriot-Watt University, Riccarton, Edinburgh, EH14 4AS; Brewing Research International, Lyttel Hall, Nutfield, Redhill, Surrey, RH1 4HY; VTT Biotechnology and Food Research, P.O. Box 1501, FIN-02044 VTT, Espoo, Finland; Université Catholique de Louvain, Unité de Brasserie et des Industries Alimentaires, Place Croix du Sud, 2 Bte 7, B-1348, Louvain-la-Neuve, Belgium; TNO Nutrition and Food Research Institute, Agro-NIBEM, P.O. Box 360, 3700 AJ Zeist, The Netherlands; TU München, Lehrstuhl für Technologie der Brauerei I and II, D-85350 Freising-Weihenstephan, Germany; Versuchs- und Lehranstalt für Brauerei, Seestrasse 13, D-13353, Berlin, Germany.

9.8   References

1. SCHILDBACH R, 'Malting barley worldwide', *Proc 27th Eur Brew Conv*, Oxford, IRL Press, 1999 299–312.

2. WILKES D, 'World malt production, usage and trade', *Ferment*, 1993 **6** 345–8.
3. ANON., *Statistical Handbook*, London, Brewing Publications, 1998 69–73.
4. BALASUBRAMANYAM V N and SALISU M A, 'Brands and the alcoholic drinks industry' in *Adding Value: Brands and Marketing in Food and Drink*, Jones G and Morgan N J (eds), London, Routledge, 1994 59–75.
5. BROCK W H, *The Fontana History of Chemistry*, London, Fontana, 1992 23–6.
6. MATHIAS P, *The Brewing Industry in England 1700–1830*, Cambridge, Cambridge University Press, 1959.
7. MACGREGOR A W, MACRI L J, BAZIN S L and SADLER G W, 'Limit dextrinase inhibitor in barley and malt and its possible role in malting and brewing', *Proc 25th Eur Brew Conv Congr*, Oxford, IRL Press, 1995 185–92.
8. ANDERSON R G, 'Exogenous enzymes in brewing', *Brewers' Guardian*, 1984 **113**(10) 15–19.
9. DICKENSON C J and ANDERSON R G, 'The relative importance of S-methylmethionine and dimethyl sulphoxide as precursors of dimethyl sulphide in beer', *Proc 18th Eur Brew Conv Congr*, Oxford, IRL Press, 1981 413–20.
10. BERRY R R, 'The physiology and microbiology of Scotch whisky production' in *Prog Ind Microbiol Vol 19*, Bushell M E W (ed.), Amsterdam, Elsevier, 1984 199–244.
11. SWAN J S, TATLOCK R R and THOMSON J, 'Sourcing oak wood for the distilling industries', *Proc 4th Aviemore Conf on Malting, Brewing and Distilling*, London, Institute of Brewing, 1994 56–70.
12. KELLY L and BRIGGS D E, 'The influence of grain microflora on the germinative physiology of barley', *J Inst Brew*, 1992 **98** 395–400.
13. GJERTSEN P, TROLLE B and ANDERSEN K, 'Gushing caused by micro-organisms, specially *Fusarium* species', *Proc 10th Eur Brew Conv Congr*, Amsterdam, Elsevier, 1965 428–38.
14. WALKER M D, BOURNE D T and WENN R V, 'The influence of malt-derived bacteria on the haze and filterability of wort and beer', *Proc 26th Eur Brew Conv Congr*, Oxford, IRL Press, 1997 191–8.
15. PALMER G H, 'Modification of abraded and normal barley grains', *J Inst Brew*, 1980 **86** 125.
16. BRIGGS D E, 'Patterns of modification in malting barley', *J Inst Brew*, 1983 **89** 260–73.
17. BROOKES P A, 'Practical experiences with barley abrasion', *The Brewer*, 1980 **66** 8–11.
18. BRIGGS D E, *Malts and Malting*, London, Blackie, 1998.
19. CANTRELL I C, ANDERSON R G and MARTIN P A, 'Grain-environment interaction in production steeping', *Proc 18th Eur Brew Conv Congr*, Oxford, IRL Press, 1981 39–46.
20. ANDERSON R G, 'Past milestones in malting, brewing and distilling', *Proc 4th Aviemore Conf on Malting, Brewing and Distilling*, London, Institute

of Brewing, 1994 5–12.

21. ESSERY R E, KIRSOP B H and POLLOCK J R A, 'Effects of water on germination tests', *J Inst Brew*, 1954 **60** 473–81.

22. BROOKES P A, 'Recent malting development', *The Brewer*, 1984 **70** 502–6.

23. WAINWRIGHT T, 'Nitrosodimethylamine: formation and palliative measures', *J Inst Brew*, 1981 **87** 264–5.

24. BAMFORTH C W, *Beer: Tap into the Art and Science of Brewing*, New York, Plenum Press, 1998.

25. LLOYD W J W, 'Adjuncts', *J Inst Brew*, 1986 **92** 336–45.

26. *Recommended Methods of Analysis of the Institute of Brewing*, 1993; *Analytica EBC*, 1987; *Methods of Analysis of the American Society of Brewing Chemists*, 1992.

27. DE BUCJ A, AERTS G, BONTE S, DUPRIE S and VAN DEN EYNDE E, 'Relationship between lipoxygenase extraction during brewing, reducing capacity of the wort and the organoleptical stability of beer', *Proc 26th Eur Brew Conv Congr*, Oxford, IRL Press, 1997 333–40.

28. SOUTH J B, 'Prediction of wort cold break performance of malt and its applications', *J Inst Brew*, 1996 **102** 149–54.

29. INATOMI K, MASUDA M and MINABE M, 'New frontiers in brewing and distilling', *Proc 4th Aviemore Conf on Malting, Brewing and Distilling*, London, Institute of Brewing, 1994 165–78.

30. RIFFKIN H L, BRINGHURST T A, McDONALD A M L and HANDS E, 'Utilisation of wheat in the Scotch whisky industry', *Ferment*, 1990 **3** 288–92.

31. TRAVIS G L, 'American bourbon whiskey production', *Ferment*, 1998 **11** 341–4.

32. WRIGHT S A, 'Canadian whisky – from grain to glass', *Ferment*, 1998 **11** 345–9.

33. SIMPSON A C, 'Advances in the spirits industry', in *Alcoholic Beverages*, Birch G G and Lindley M G (eds), London, Elsevier, 1985 51–67.

34. PYKE M, 'The manufacture of Scotch grain whisky', *J Inst Brew,* 1965 **71** 209–18.

35. PALMER G H, 'Scientific review of Scotch malt whisky', *Ferment*, 1997 **10** 367–79.

36. WAINWRIGHT T, 'New Glenfiddich whiskies', *Ferment*, 1998 **11** 335–9.

37. WALKER M W, 'By-products – distilling', *Ferment*, 1988 **1**(2) 45–7.

38. HAMMOND J R M, 'Research and Development', *EBC Good Practice Manual*, Nürnberg, Getränke-Fachverlag Hans Carl, 1998.

39. ENARI T-M, 'State of the art of brewing research', *Proc 25th Eur Brew Conv Cong*, Oxford, IRL Press, 1995 1–11.

40. JONES G and MORGAN N J (eds), *Adding Value: Brands and Marketing in Food and Drink*, London, Routledge, 1994.

41. DEVREUX A, 'The 27th Congress of the EBC at Cannes 29 May–3 June 1999', *J Inst Brew*, 1999 **105** 129–37.

42. GORDON R W (Scottish Courage), Personal Communication, 29 March 1999.

43. BOCK K (Carlsberg), Personal Communication, 9 April 1999.
44. VAN DEN HOUTEN G J, (South African Breweries), Personal Communication, 21 May 1999.
45. OUTTRUP H and PETERSEN S G, 'PCR screening methods for transgenic raw materials', *Search and Research*, Copenhagen, Carlsberg Research Centre, 1997 10–11.
46. VAN DUIJN G, BLEEKER-MARCELIS H and HESSING M, 'Detection methods for genetically modified crops', *Proc 27th Eur Brew Conv Congr*, Oxford, IRL Press, 1999, 509–13.
47. CAPPELEN S, 'Carlsberg er bange for gensplejsede majs', *Ingeniøren*, 12 March 1999.
48. PALMER G H, 'Scientific review of Scotch malt whisky', *Ferment*, 1997 **10** 367–79.
49. HAMMOND J R M and BAMFORTH C W, 'Genetic engineering in brewing', *Biotechnol Genet Eng Rev*, 1993 **11** 147–68.
50. BAXTER D, 'The application of genetics to brewing', *Ferment*, 1995 **8** 307–15.
51. HOLM P, KNUDSEN S, OLSEN J, JENSEN J and MÜLLER M, 'Genetic transformation of barley for quality trials', *Ferment*, 1992 **5** 417–24.
52. RIGHELATO R, 'Technology foresight: alcoholic drinks sector report', *Ferment*, 1998 **11** 93–5.
53. MANNONEN L, KURTÉN U, RITALA A, SALMENKALLIO-MARTTILA M, HANNUS R, ASPEGREN K, TEERI T and KAUPPINEN V, 'Biotechnology for the improvement of malting barley', *Proc 24th Eur Brew Conv Congr*, Oxford, IRL Press, 1993 85–93.
54. KIHARA M, OKADO Y, KURODA H, SAEKI K, YOSHIGI N and ITO K, 'Generation of fertile transgenic barley synthesizing thermostable β-amylase', *Proc 26th Eur Brew Conv Congr*, Oxford, IRL Press, 1997 83–90.
55. WAN Y and LEMAUX P, 'Generation of large numbers of independently transformed fertile barley plants', *Plant Physiol*, 1994 **104** 37–48.
56. JANSEN L G, OLSEN O, KOPS O, WOLF N, THOMSEN K K and VON WETTSTEIN D, 'Transgenic barley expressing a protein-engineered thermostable (1-3,1-4)-β-glucanase during germination', *Proc Natl Acad Sci, USA*, 1996 **93** 3487–91.
57. MANNONEN L, RITALA A, NUUTILA A M, KURTÉN U, ASPEGREN K, TEERI T H, AIKASALO R, TAMMISOLA J and KAUPPINEN V, 'Thermotolerant fungal glucanase in malting barley', *Proc 26th Eur Brew Conv Congr*, Oxford, IRL Press, 1997 91–100.
58. JENDE-STRID B, 'Proanthocyanidin-free malting barley – a solution of the beer haze problem', *Proc 26th Eur Brew Conv Congr*, Oxford, IRL Press, 1997 101–8.
59. BUTLER D and REICHHARDT T, 'Long-term effect of GM crops serves up food for thought', *Nature*, 1999 **398** 651–6.
60. McKIE R and ARLIDGE J, 'Gene food scientists reap a harvest of bitterness', *The Observer*, 23 May 1999 12–13.
61. REDFORD T, 'European sceptics resist GM foods. Less concern in US about

scientific advances', *The Guardian*, 16 July 1999.

62. ATKINSON B N, 'Biotechnology in the brewing industry', *Proc 19th Eur Brew Conv Congr*, Oxford, IRL Press, 1983 339–51.

63. MOLZAHN S W, 'Genetic engineering: yeast', *Proc 21st Eur Brew Conv Congr*, Oxford, IRL Press, 1987 197–208.

64. HAMMOND J R M, 'The development of brewing processes: the impact of European biotechnology regulations', *Proc 23rd Eur Brew Conv Congr*, Oxford, IRL Press, 1991 393–400.

65. ANDERSON R G, 'The pattern of brewing research: a personal view of the history of brewing chemistry in the British Isles', *J Inst Brew*, 1992, **98** 85–109.

66. BRIGGS D E, *Barley*, London, Chapman and Hall, 1978.

67. PALMER G H, *Cereal Science and Technology*, Aberdeen, Aberdeen University Press, 1989.

68. HOUGH J S, BRIGGS D E, STEVENS R and YOUNG T W, *Malting and Brewing Science*, London, Chapman and Hall, 1982.

69. YOUNG T W and LEWIS M J, *Brewing*, London, Chapman and Hall, 1995.

70. HARDWICK W, *Handbook of Brewing*, New York, Marcel Dekker, 1995.

71. KUNZE W, *Technology: Brewing and Malting*, 7th edition translated by T Wainwright, Berlin, VLB, 1996.

72. PIGGOTT J R, SHARP R and DUNCAN R E, *The Science and Technology of Whiskies*, Harlow, Longman, 1989.

10

Current practice in cereal production

E. J. Evans, University of Newcastle

10.1   Introduction

The dominant cereals of the United Kingdom are wheat and barley; oats, once
the most widely grown grain crop, has declined to a minor cereal during the last
sixty years. Rye, whilst a major cereal in mainland Europe, is relatively
unimportant in UK agriculture. Similarly small areas of triticale, a hybrid of
wheat and rye, and mixed corn are grown for livestock feeding. The importance
of cereals has long been recognised in both world and UK agriculture.
Advances in plant breeding and the adoption of highly efficient production
systems have combined to bring about almost a fourfold increase in grain yield
during the second half of the twentieth century. This initially secured the
profitability of arable farming in lowland Britain, but with continued high levels
of production across western Europe grain surpluses have become a major
burden on EU finances. Measures adopted within the CAP have resulted in a
sharp decline in the overall profitability of cereal production during the latter
half of the 1990s.

10.1.1   Trends in cereal production

The total area of land used for the cultivation of cereals in Britain has increased
from under three million hectares to over seven million hectares during the
twentieth century. Furthermore, during this period there was a significant change
in the relative importance of wheat, barley and oats (Fig. 10.1).

The dominant cereal during the period 1898 to 1938 was the oat crop, grown
predominantly to satisfy the dietary needs of the farm horse as the principal
source of power on the arable farm. At this time barley was largely confined to

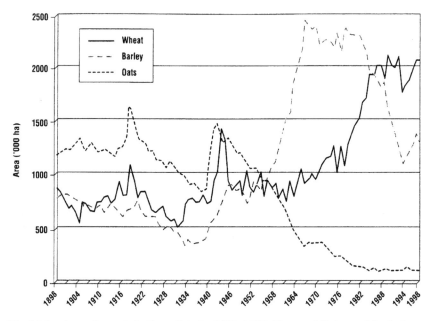

Fig. 10.1   Area of cereals in Great Britain, 1898–1998. (Source: Ministry of Agriculture, Fisheries and Food.)

the production of beer and spirits, whilst a greater reliance was placed on imported rather than home produced wheat for human consumption.

With the outbreak of the Second World War the need to reduce reliance on imported North American wheat resulted in a major Government initiative to promote cereal production in Britain. This 'ploughing out' campaign was successful in increasing the cultivation of wheat and oats on land which hitherto had been considered only marginal for cereal production. Following the end of hostilities the area of wheat and barley remained above those of the pre-war period, due in part to the introduction of a guaranteed price support mechanism for cereals, and the introduction of appropriate machinery to assist cultivation and crop establishment. As the tractor replaced the farm horse the demand for feed oats declined, a trend which has continued to the present time.

During the 1950s barley became the most important cereal crop after it was demonstrated that the grain could be satisfactorily fed to farm animals under more intensive systems of production. In parts of southern England continuous spring barley growing was practised with the adoption of increased mechanisation on large fields made possible through the removal of hedges and ditches. This trend continued until the mid 1960s when the profitability of continuous barley began to decline as production costs began to escalate and the agronomic limitations of continuous cereal production became more apparent.

Between 1945 and 1970 the area of wheat remained fairly static, year to year variation being largely accounted for by the climate during the autumn. Unlike barley most wheat crops at that time were autumn sown; consequently in a wet autumn the area sown to wheat would be lower than in a dry autumn, although other factors, such as the need to control weeds and soil-borne diseases, also had an impact.

With Britain's entry into the EEC the profitability of wheat increased significantly compared to that of barley with the result that the area of barley began to decline and that of wheat increase. In the early 1980s wheat became the dominant UK cereal. Higher wheat prices, although important, was only one of several factors responsible for the increasing popularity of wheat. The introduction of high yielding winter varieties, the adoption of effective fungicide and herbicide programmes, and the availability of plant growth regulators all combined to achieve high yields and satisfactory financial returns. Improved grain quality and a better appreciation of market requirements also made a significant contribution.

10.1.2   Cereal yields

Cereal yields changed very little during the first half of the twentieth century, but have more than trebled since. There are a number of factors which have contributed to this trend. The rediscovery of Mendel's work provided the scientific basis on which cereal breeding could develop, first within the public sector and more recently by private companies which have now combined into multinational conglomerates. Legislation at national and European levels enabled plant breeding companies to recover their costs in the form of royalties which enabled further advances to be made in developing high yielding cereal varieties with improved quality characteristics. Future developments in cereal breeding will become increasingly dependent on advances in biotechnology and the willingness of the public to accept genetic modification into the food chain.

Austin (1978) calculated the potential yield of winter wheat to be in the region of 13 t/ha. Some high yielding crops have achieved this level of performance, but of more significance has been the constant upward trend in the national average yield (Fig. 10.2).

Although all cereals have benefited greatly from these advances in plant breeding, winter wheat yields have improved to a greater extent than those of oats and barley. During the period 1947 to 1978, Silvey (1981) estimated that the improvement in the national average yields of wheat, barley and oats was of the order of 105%, 76% and 87% respectively. The contribution made by the adoption of new varieties was considered to account for approximately 50% of this improvement.

New varieties undergo extensive field trials to ascertain their field performance and quality characteristics prior to their wide scale commercial adoption. Cereal producers must purchase certified seed or use seed that has been saved from the previous year's crop from their own farm as 'home saved seed'. This ensures high seed quality with good germination capacity, absence of impurities and free from seed borne diseases. Thus advances in plant breeding are rapidly transmitted into farming practice.

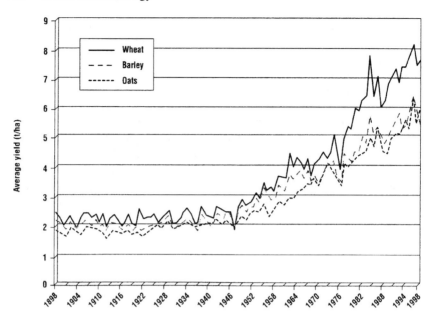

Fig. 10.2 Average yield of cereals in Great Britain, 1898–1998. (Source: Ministry of Agriculture, Fisheries and Food.)

10.1.3 UK regional distribution of the major cereals

Figures 10.3 and 10.4 show the UK distribution of wheat and barley respectively. Traditionally wheat was associated with the heavy soils of eastern England and the East Midlands, but has increasingly been cultivated throughout the eastern side of the country, frequently on lighter textured soils. Winter wheat requires a combination of adequate sunshine, especially during the grain filling stage through to final harvest, coupled with an adequate supply of soil moisture. These requirements become even more important for the production of high quality grain for human consumption.

Barley is less demanding than wheat both in terms\ of its soil and climatic requirements. The crop is frequently associated with light land, chalk and limestone soils in the south and east of England. In western and northern regions of the UK the crop is mainly grown as a livestock feed, although the highest quality malting barley is frequently produced on the light textured soils in eastern England and Scotland.

Oats are currently cultivated in relatively few areas of the UK. The crop remains popular in Scotland, the south west, Wales and the Welsh border counties, although the best quality grain for human consumption is produced in eastern England.

Fig. 10.3 Regional distribution of wheat in the UK, 1998. (Source: Ministry of Agriculture, Fisheries and Food.)

10.1.4 Grain quality and market outlets

There are two main outlets for grain, animal feed, and human consumption, while a small amount is required annually for seed; grain is also exported from the UK, mainly to Europe (Table 10.1). The requirements of these different markets vary considerably, but are essentially defined by a number of 'quality' attributes, which determine the suitability of a sample of grain for a particular end use. The definition of quality therefore depends on the requirements of the specific market.

Grain attributes which determine its suitability for a specific market include its chemical, physical and biological properties. All sectors of the market have a basic requirement for sound grain free from impurities, insect damage and moulds. Other standards are more market specific and will vary in importance according to species and end product. For wheat these may include protein quality and quantity, endosperm texture, flour yield and colour, water adsorption capacity, α-amylase activity and specific weight. For barley appearance, varietal purity, seed vigour and germination, specific weight and moisture content are generally more important, whilst for oats sound well developed grain with a high kernel content is an important determinant of quality.

A number of these attributes are genetically controlled, others are dependent on crop management during both the growing and storage periods, whilst

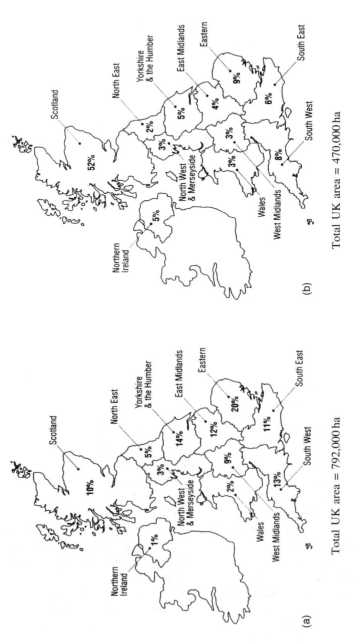

Fig. 10.4 Regional distribution of (a) winter barley and (b) spring barley in the UK, 1998. (Source: Ministry of Agriculture, Fisheries and Food.)

(a)

Total UK area = 792,000 ha

Scotland 10%

North East

Yorkshire
& the Humber 5%

East Midlands 14%

Eastern 20%

Northern Ireland 1%

North West
& Merseyside 3%

Wales 2%

West Midlands 9%

South West 13%

South East 11%

(b)

Total UK area = 470,000 ha

Scotland 52%

North East

Yorkshire
& the Humber 5%

East Midlands 4%

Eastern 9%

Northern Ireland 5%

North West
& Merseyside 3%

Wales 3%

West Midlands 3%

South West 8%

South East 6%

Table 10.1 Annual UK cereal supply and demand estimates, 1998/9 ('000 tonnes)

	Wheat	Barley	Oats
Opening stocks	1,358	2,013	33
Production	15,104	6,632	587
Imports	1,124	162	8
Total availability	17,946	8,807	628
Human and industrial use	6,372	1,950	270
Home grown	(5,390)	(1,842)	(262)
Animal feed	6,336	3,185	255
Seed	337	200	18
Other	77	33	3
Total consumption	13,122	5,368	546
Exports and intervention	3,348	2,590	0

Source: Home Grown Cereals Authority, Market Information (mi), 20 December 1999.

climatic conditions during grain filling and harvesting can often determine overall quality standard. Hence year to year variation in grain quality will have a major influence on the balance between supply and demand, which in turn will influence the premium paid to growers over and above that paid for feed grain.

Currently some six million tonnes of wheat and three million tonnes of barley are used as animal feed in the UK each year. During the pre-war period oats was the dominant feed grain, used largely to feed the farm horse, the primary source of power on the arable farm. In the 1960s barley became the dominant feed grain, but more recently wheat with its lower fibre and higher energy and protein content has been used increasingly by the compound feed manufacturing industry. Quality criteria for feed grain have not been well established; consequently very low priority is given to breeding programmes to improve feeding characteristics, although low specific weight grain is considered to be of lower nutritional value.

The major human and industrial uses of wheat in the UK are for breadmaking, biscuit manufacture and distilling. Each year approximately five million tonnes of wheat is milled into flour; two thirds of this is used for breadmaking, the remainder is used in the manufacture of biscuits and other food products. The introduction of the Chorleywood Breadmaking Process has allowed the use of lower protein flour which has enabled a much higher inclusion of home-grown wheat in the manufacture of the standard white loaf. Although most biscuit flour is made from home-grown wheat, some products such as wafers and crackers may require specific varieties or the addition of gluten modifying additives to achieve the desired eating qualities.

Specific weight, a measure of the bulk density of grain, is widely used as a wheat quality indicator. High specific weight grain results in better flour extraction within a specific variety, but is not always consistent across different varieties. It is a crude measure of grain fill and can vary from 40 kg hl to over

80 kg hl for plump, well filled grain. Breadmaking samples are normally over 76 kg hl. Seasonal variation in specific weights of individual varieties is quite marked, largely on account of differences in radiation and rainfall during the grain filling period. However, management practices are also important in ensuring that the plant remains free of pests and disease and is supplied with adequate nutrients and water.

Hard wheat is required for inclusion in breadmaking and a soft texture is required for biscuit making. When hard wheat is milled the endosperm cells separate along the cell wall margins into easily sifted particles, and the bran is easily separated from the endosperm to give a high extraction of white flour, capable of high water adsorption during dough production. Soft wheat has a much lower extraction rate and the flour consists of a mass of fine cell debris with poor flow characteristics and a lower water adsorption capacity. Endosperm texture is under genetic control with varieties either classed as hard or soft. This character is simply inherited and easily managed within breeding programmes, whilst agronomic practices have no influence on this quality component.

Wheat flour is used for breadmaking as a result of the viscoelastic properties of the dough when water is added. The dough may be classed as either strong or weak, depending on the quantity and quality of the grain proteins, which in turn influences gluten strength. For breadmaking, gluten must be strong enough to retain the carbon dioxide generated during fermentation, allowing the bread to rise. The protein content of wheat grain varies widely, but for breadmaking a value of at least 11% is required. In practice high grain protein levels are achieved through the application of nitrogen fertiliser above the optimum for yield. This is especially beneficial when applied late in the season. Foliar applications of urea applied during the milk development stage have been found to be particularly beneficial. Recent work has also demonstrated the importance of supplying the crop with an adequate supply of sulphur at a time when atmospheric depositions have declined significantly as a result of reduced pollution. Applications of around 20 kg S ha in spring have been shown to improve the breadmaking characteristics of wheat grown in sulphur-deficient areas of the UK.

Protein quality is strongly influenced by genotype, although husbandry and environmental factors can also play an important role. Low grain sulphur will result in low concentrations of the sulphur-containing amino acids, cystine and methionine, and may result in poor loaf volume. Protein quality has also been shown to fall as a result of late fungicide sprays that prolong the grain filling period.

During seed germination endosperm starch is converted into soluble glucose and maltose to support the developing embryo. This is brought about by enzymatic activity, especially the enzyme α-amylase, present within the grain and activated during the germination process. Some α-amylase activity is needed to release sugars and aid fermentation during the breadmaking process. Excessive α-amylase levels result in the formation of a darkened loaf crust as a result of sugar caramelization and a sticky crumb structure which can cause problems during slicing. Genetic variation exists in the amount of α-amylase activity both during pre-maturity and enzyme formation during post dormancy sprouting. These two components are inherited independently giving rise to a situation where varieties differ in their α-amylase content where there is no visible sprouting. Varieties with low α-amylase activity combined with good resistance to sprouting are favoured for breadmaking.

Grain α-amylase levels reach their lowest levels during ripening, thereafter increasing significantly. This suggests an optimum date for harvesting, but one which is difficult to predict and achieve in practice. Nevertheless, it is good practice to harvest crops destined for the breadmaking market early to avoid the effects of wet weather.

Post-harvest management is an essential part in achieving high quality grain to meet specific market requirements. Food hygiene regulations apply to stored grain which must be protected from moulds, bacteria, rodent, bird and insect damage, whilst pesticide residues must not exceed UK statutory levels. These standards form the basis of recently introduced quality assurance schemes, combining all facets of grain production from field to store and on to the mill. Breadmaking grain harvested at moisture contents in excess of 15% must be dried carefully at air temperatures below 60°C to avoid denaturing the proteins, thereby destroying the elasticity of the gluten.

For biscuit production, flour is produced from soft milling varieties with high extensibility to ensure that the different biscuit shapes cut from the dough retain their outline after cutting. Wheat with a protein level below 10% is preferred to reduce gluten elasticity. The low water absorption characteristics of soft endosperm wheat are desirable to limit the need for drying the final product to a standard moisture content and also to reduce cracking during cooling and subsequent packaging and storage. The α-amylase content of the flour is also less critical than it is for breadmaking.

For distilling, soft endosperm texture is required and whilst there is no protein specification a high protein content can lead to problems with low starch and spirit yields. Distillers often prefer to select specific soft wheat varieties that have consistently produced high yields of spirit.

Grain quality has an extremely important influence on the suitability of a sample of barley grain for malting. Maltsters require grain from a recognised malting barley variety which is of uniform size with low husk and nitrogen content with a high germination capacity. These malting characteristics can only be accurately measured with a micro malting test which is both slow and expensive, although the physical condition of the grain and its nitrogen content are generally considered to be a good guide to its malting potential.

Traditionally spring varieties have commanded the highest malting premiums, but with the increased popularity of winter compared to spring barley in the UK, plant breeders have produced a number of good winter malting barley varieties. All new varieties are evaluated for their malting characteristics from micro malting tests to assess their hot water extract and are graded accordingly. However, not all high graded varieties are approved by the Institute of Brewing until they have been subjected to commercial scale malting. Two-row barley varieties are preferred to six-row varieties because of the evenness of the grain, whilst varietal admixtures are not acceptable for malting.

The characteristics of a good malting variety are that it readily takes up water on steeping, germinates readily and evenly and produces high levels of hydrolytic enzymes for the conversion of the starch to soluble sugars. The malting process is effectively a process of controlled germination and every effort must therefore be made to retain a high germination capacity. Poor germination can result from careless threshing damaging the embryo, from drying at too high a temperature, or from heating caused by storage at too high a moisture content. Furthermore, uniform, complete germination can only be achieved by avoiding crop lodging and harvesting the grain when it is fully ripe. The physical appearance of the grain is also important; maltsters prefer samples where the grain is uniformly plump. This is usually measured by passing the grain through a sieve and determining the percentage grain retained on the sieve. Screening standards vary slightly between England and Scotland. Samples should also have a low husk and high endosperm content, free from broken grains and damaged husk.

There is a very good correlation between grain nitrogen and the amount of malt extract achieved, low grain nitrogen giving more fermentable extract. During the past decade the swing from traditional cask conditioned draught ale towards light lager beers has reduced the demand for very low nitrogen grain. For the former a nitrogen content of 1.5 to 1.65% is required whereas for lager beer a grain nitrogen content of 1.8% is acceptable. Traditionally the lowest grain nitrogen samples have been obtained from crops grown on light sandy soils along the eastern coast of Britain. These soils have low residual levels of organic nitrogen which allows better control of plant nitrogen supply from annual fertiliser nitrogen applications. The amount of nitrogen applied to a malting barley crop will be less than that applied to a crop destined for the feed market, thereby sacrificing some yield potential for low nitrogen grain. Applications of spring nitrogen to both the winter and spring sown crops should also be completed at an early stage of crop development.

About a half of the UK oat crop is used for human consumption, whilst the remainder is used for animal feeding. Food products containing oats include oatmeal for porridge, oatcake, muesli and other breakfast cereal products. In many of these products it is necessary to remove the fibrous husk surrounding the kernel mechanically, although in recent years the introduction of naked or huskless oats has been a major breeding achievement. Husk content has also been found to differ between varieties and consequently some varieties are more likely to attract a premium.

Traditionally oats were regarded as a low input crop; however, if a high quality product is required for human consumption then more careful management of inputs is required. Foremost amongst these is the need to avoid the crop lodging with careful attention to the time and rate of nitrogen application, coupled with the use of plant growth regulators. Maintaining the crop free from weeds and diseases, especially mildew, is important in achieving a high yield of quality grain.

10.2   Varietal selection

The contribution of new cereal varieties to the profitability of grain production is well recognised. The prime objective of plant breeders from the outset has been to develop high yielding varieties, with good disease resistance against the major cereal pathogens. This continues to be a major aim of current breeding programmes, although increased attention is now being directed towards improving grain quality to meet specific market requirements.

In wheat, and to a lesser extent barley, improved grain yields have been achieved through selection of short strawed varieties. Old and new varieties of cereals differ very little in the amount of total dry matter accumulated during the growing season, but selection for a higher Harvest Index (HI) has resulted in a greater proportion of this biomass being accumulated in the grain. The HI of varieties in cultivation during the period 1930–40 was around 30%, compared to between 50 and 55% for present day varieties (Austin *et al*. 1980). Further agronomic advances have been achieved with selection for improved standing power, whilst increased resistance to grain shedding and sprouting have contributed significantly to advances in wheat breeding.

Breeding new varieties with improved resistance to the major cereal diseases has always been a key objective for plant breeders. Before the introduction of broad spectrum systemic fungicides in the late 1970s, genetic resistance was often the only defence mechanism available to growers and was often the most important factor to be considered when choosing a variety. Good varietal resistance remains an important consideration, even within current husbandry systems in which routine applications of two or three fungicides is commonplace during the growing season. Poor genetic resistance can lead to a rapid epidemic build up of disease, which may prove difficult and expensive to control leading to a detrimental effect on yield and quality.

The suitability of wheat or barley for different end uses is partly under genetic control, although varietal differences can be modified by husbandry and weather conditions. Characteristics such as the endosperm texture of wheat and the malting potential of barley is entirely under genetic control, while quality parameters such as specific weight are influenced to a greater extent by husbandry and climate. Considerable advances have been made in developing varieties with improved quality attributes through a better understanding of their molecular basis. At the same time market needs have become more clearly defined which is likely to lead to better targeting of new varieties for specific markets in the future.

New varieties must demonstrate that they are distinct in their genetic make-up, uniform and stable in their characteristics and have value for cultivation and use before they are added to the UK National List. Further evaluation with and without fungicides in trials throughout the country provides further information on their yield and quality characteristics. Recommendation depends on their average performance exceeding the mean performance of varieties already recommended.

10.3   Crop establishment

Traditionally the UK wheat crop has been predominantly winter sown, while barley was largely spring sown until the mid 1970s when the winter crop increased in popularity through the introduction of high yielding feed varieties. Spring barley varieties remain the dominant malting types. In England and Wales winter oats are preferred, whereas in Scotland spring oats dominate.

The optimum sowing time for both winter wheat and barley is from mid September to early October, although with the increase in farm size and the reduction in full-time farm labour both wheat and barley are now sown earlier in the autumn. Early sowing is beneficial in promoting good root and shoot systems before the onset of winter, which in turn enables the crop to intercept a greater proportion of available radiation during spring and summer, thereby establishing a higher yield potential. To achieve this potential growers frequently incur additional production costs, especially through increased use of herbicides and fungicides.

Spring cereals are normally sown as soon as soil conditions allow from February onwards, although spring varieties of wheat may be sown during the period from the middle of November until January after harvesting sugar beet and potatoes on light land. The yield potential of late sown spring wheat and barley is generally inferior to crops sown before the end of March.

Soil conditions during seedbed preparation have a significant effect on crop performance. Seedbeds have normally been produced by ploughing, followed by a series of secondary cultivations to produce a satisfactory tilth. Whilst effective, this method of crop establishment is both time consuming and expensive. Direct drilling of winter cereals became popular in the 1970s as a cost effective system of establishment, but the introduction of the straw burning ban in England and Wales, coupled with an increase in grass weeds has made it less effective in recent times. Minimum cultivation systems have become a popular method of crop establishment on suitable soils in the absence of major grass weed problems. Recent advances in machine design, combining cultivation and drilling machinery behind a single power unit, have significantly increased the ability of growers to sow large areas quickly and effectively. Reducing establishment costs is now regarded as a major factor in containing fixed costs.

An important feature of all cereal plants is their ability to produce a large number of shoots or tillers, although by harvest many of the tillers will have died, leaving a main stem and one to three ear bearing tillers. The number of tillers produced is often negatively correlated with plant density, while tiller survival is strongly influenced by environmental and agronomic factors which, in turn, influence assimilate supply. The rate and timing of nitrogen can have an important role in both the production and survival of tillers; indeed it is one of the main tools that the grower has at his disposal to influence ear population at final harvest. The practical significance of this is that satisfactory yields can be achieved from a wide range of sowing rates, especially for winter wheat and barley.

The target population of established plants is between 200 and 300 plants/m^2. To achieve this between 100 and 200 kg/ha of seed is sown, depending on seed size. The higher the individual seed weight, the higher the seed rate required to achieve the target plant population. Allowance must also be made for seedling mortality and the loss of established plants to pest and disease attack. The use of fungicide seed dressings is common practice to control soil and seed borne pathogens. Seedling losses can be higher in winter compared to spring sown cereals.

Cereals are relatively insensitive to variations in row width and inter-row spacing. Similarly sowing depth is less critical than in many of the smaller seeded arable crops, such as oilseed rape. Sowing at the target depth of 2 to 3 cm is easily achieved with conventional drills.

10.4   Crop nutrition

The yield and quality of all cereal crops is strongly dependent on the availability of an adequate supply of soil mineral nutrients throughout the growing season. The higher the yield potential the higher the nutrient demand, while the grain nitrogen content is a major quality determinant in wheat and barley.

The nutrient status of most arable soils in the UK is too low to achieve satisfactory yields in the absence of applied nutrients. Crop demand is normally met through the application of inorganic fertilisers, although in organic cereal systems additional nutrients can only be supplied from manures and other organic sources. The results of numerous field trials over the last fifty years have provided a sound basis on which to base the nutrient requirements of cereals. Compared to many arable crops, cereals have a relatively low demand for phosphate and potassium. As a rule of thumb phosphate and potassium are applied at the rate of 10 kg/tonne of expected grain yield to replace the P and K removed and maintain soil reserves. The amount applied may be reduced slightly in soils with high P and K reserves. In recent years with the decline in atmospheric SO_2 deposition and the trend away from sulphur-containing fertilisers some cereal crops grown on light land may benefit from sulphur application. Recent research has also identified the importance of sulphur to the breadmaking qualities of wheat.

Cereals do not have a high demand for trace elements, although copper and manganese deficiencies have been recorded in wheat and barley. Copper deficiency is frequently associated with high organic matter soils; symptoms of deficiency are most common when the ears emerge as white heads, producing shrivelled grain with low specific weight. Manganese deficiency is most frequently observed on light sandy, or chalk soils with high pH values. Low availability of manganese results in grey-white lesions on the leaves. Deficiencies of other trace elements are less common, although it is desirable to monitor the trace element content of crops from time to time through plant analysis. Maintaining the soil pH at an appropriate level will also be beneficial.

Nitrogen is by far the most important nutrient, influencing both grain yield and quality. With the exception of the high organic peat soils the levels of available soil nitrogen in most arable soils is well below that which is required for satisfactory growth and high yields of grain. Total soil nitrogen, on the other hand, is often present well in excess of crop requirements but is present in the organic matter fraction and is therefore dependent on microbial activity to release the nitrogen into a mineral form, suitable for crop uptake. This process of mineralisation is very dependent on soil environmental conditions, particularly an adequate level of soil moisture, high temperature and aerobic soil conditions. The release of nitrogen from this organic fraction will progress at a faster rate in spring and autumn than at other times of the year. Nevertheless, it is difficult to predict accurately rates of mineralisation throughout the growing season in such a way that it would be possible to adjust fertiliser inputs accurately to meet the precise nitrogen demand of crops. A significant proportion of the mineral nitrogen fraction not used by the crop is leached into water courses, presenting a potential environmental hazard.

The role of nitrogen in promoting grain yield has often been evaluated in terms of its effect on single plant components such as leaf area, tiller production and survival, grain weight and number. More recently, research efforts have been directed at evaluating the effects of nitrogen on the whole green crop canopy, in particular its role in manipulating the canopy size to enable maximum radiation interception over as long a time interval as possible. This approach, which has become known as canopy management, is currently being evaluated in commercial wheat crops throughout the UK (Sylvester-Bradley et al. 2000).

The amount of inorganic nitrogen fertiliser applied to cereals in the UK has increased appreciably over the last thirty years. There have been a number of agronomic reasons for this trend, but without question the driving force has been the realisation by cereal producers that high nitrogen rates are associated with high yields and hence higher financial returns. This has been made possible through the introduction of high yielding varieties with improved lodging and disease resistance, coupled with the adoption of highly effective fungicide programmes to control foliar diseases and plant growth regulators to improve further the crop's standing ability.

The nitrogen requirements of winter cereals are significantly higher than those of spring sown cereals on account of their longer growing season and higher demand for biomass production. Additionally, the timing of N application will vary according to their very different growth patterns. The results of numerous field trials have clearly demonstrated that there is very little benefit from the application of autumn nitrogen to winter wheat or barley. Indeed there may be an excess of nitrogen in seedbeds prepared by traditional ploughing arising from the breakdown of organic matter. Applications of inorganic nitrogen at this time only serves to increase the likelihood of nitrate leaching. Winter cereals established by direct drilling have been shown to benefit from a moderate application of nitrogen in the autumn.

The amount of spring and summer nitrogen required for optimum yields is lower in winter barley than winter wheat, largely due to the former being more prone to lodging than wheat. The rates of application can vary from as little as 50 kg/ha to over 250 kg/ha, depending on a range of agronomic and environmental factors; the most important being the previous cropping, soil type, expected grain yield and overwinter rainfall, while other variables such as the variety's standing ability, disease resistance and crop density must also be taken into account.

The effectiveness of applied nitrogen in promoting crop growth and ultimately grain yield is greatly improved by ensuring that adequate nitrogen is made available according to the demands of the crop. If this objective is achieved then the loss of soil nitrates through leaching will also be minimised. Normal practice is to apply the spring nitrogen in two dressings, the first in late February to early March and the second some time during the period late March to late April. The first of these dressings, of approximately 40 kg/ha, will be applied according to ground conditions and the likely demand for nitrogen by the crop. The second or main application is generally considered to have the greater effect on grain yield and is applied to coincide with the start of stem extension. The window of application will therefore vary according to sowing date, variety and spring temperatures. The timing of the main spring nitrogen to winter sown malting barley is critical in achieving low grain nitrogen concentrations and should be completed in March, irrespective of the crop's stage of growth.

Winter wheat grown for breadmaking requires an additional application of late nitrogen to promote grain protein content. This can either be applied as a solid fertiliser a few weeks after the main application, or as an urea-based solution sprayed onto the crop during grain filling.

The demand for nitrogen by spring cereals is considerably lower than for winter cereals due to the short growing season leading to reduced biomass production. The risk of nitrate leaching is also greatly reduced and therefore the benefits of splitting the application is considerably less than for winter cereals, although with high rates of nitrogen in excess of 125 kg/ha, there may be some practical benefits in splitting the application. To produce low grain nitrogen spring barley it is desirable to apply all the nitrogen before the two leaf stage. In contrast spring wheat for breadmaking would benefit from a late season application of nitrogen to promote grain protein levels.

10.5   Weed control

The presence of weeds will reduce the yield of all cereal crops due to their ability to compete with the crop for water and nutrients from the soil and intercept radiation that would otherwise be utilised by the crop canopy. The combined effects of weed competition on crop yield is greater during the early stages of growth; therefore adequate weed control measures are essential during crop establishment. Some weed species can also have an indirect effect on grain yield by increasing the likelihood of lodging and encouraging the development of pests and diseases. The presence of weeds during the later stages of grain growth is likely to have a detrimental effect on grain quality and often interfere with the harvesting operations.

Intensive cereal growing systems rely heavily on the use of chemical weed control programmes based on the use of one or more products applied either pre- or post-emergence of the cereal crop. At the other end of the spectrum organic cereal production systems are based on non-chemical control measures. Here the options are limited to the use of mechanical weeders and the manipulation of crop growth to allow it to become dominant and to out-compete the weeds. In future it is likely that attention will be given to a more integrated approach to weed control with strategic use of herbicides in combination with cultivation practices. The preparation of a stale seed bed through shallow cultivations to encourage germination of weeds which are then destroyed before the crop is sown is one such technique which may be reintroduced in an attempt to reduce the cost of chemical weed control. Further attention will also be directed to considering the weed spectrum across the whole rotation rather than in a piecemeal, crop by crop basis.

Herbicides represent between 35% and 40% of crop protection costs for winter sown cereals in the UK. Expenditure on weed control will often vary from field to field depending on the weed spectrum present. For example, the cost of controlling grass weeds will be a good deal more expensive than the cost of controlling broad-leaved weeds such as poppy or shepherd's purse. Selection of the appropriate herbicide programme is dependent in the first instance on being able to identify the weed flora present, recognising the dominant weed species within this spectrum and having an appreciation of the likely size of the population. A number of attempts have been made to identify weed thresholds, below which treatment may be deemed unnecessary, but these have been difficult to apply in practice due to the different competitive characteristics of different weed species and their effects on crop growth.

An effective weed control strategy is also dependent on understanding the mode of action of the active ingredient, its interaction with other chemicals with which it may be mixed and its effect on the target weed species. In general herbicides are more effective against weeds during their early stages of growth. Rates of application are determined by the manufacturers, based on many seasons of field and glasshouse evaluation trials, although in an attempt to implement more cost effective weed control strategies growers often adopt reduced dose rates according to prevailing crop and weather conditions.

It is now well recognised that in many areas of southern and eastern England populations of blackgrass (*Alopecurus myosuroides*) have developed that are resistant to the most commonly used herbicides for grass weed control. Resistance has been mainly associated with continuous, or near continuous winter cereal cropping, often established by non-ploughing techniques and regular use of a narrow range of herbicides to control grass weeds. In general grass weeds are more expensive and more difficult to control than broad-leaved weeds.

10.6   Disease control

Cereals are prone to a range of diseases caused by micro-organisms, predominantly fungi, which can attack the roots, stems, foliage and/or the ear, causing substantial losses of yield and frequently having a detrimental effect on grain quality. The presence of disease in an otherwise healthy crop is first recognised in the field by the appearance of well defined symptoms resulting from earlier activity on the part of the pathogen. If the disease is allowed to progress then the pathogen itself becomes more obvious, but often more difficult to control at this stage of its development.

Disease control has traditionally been based on cultural practices aimed at improving the ability of the crop to resist infection or attempting to interfere with the life cycle of the pathogen. Exploiting genetic resistance through careful selection of varieties remains an important disease control strategy but over the last thirty years fungicides have become an integral part of cereal production systems in the UK.

Conditions have been identified from numerous field trials that promote the establishment and development of pathogens on cereal crops. For winter sown cereals, early sowing and a high soil nitrogen status in the autumn promote lush soft tissue which can become very prone to a number of the leaf infecting diseases such as mildew (*Erysiphe graminis*) and brown and yellow rust (*Puccinia* spp.) with mild autumn weather. The frequency with which cereal crops are grown in the rotation can also be a major determinant of crop susceptibility to a range of pathogens. The most important disease in this respect is undoubtedly take-all (*Gaeumannomyces gramminis*) which is particularly serious in winter wheat in second and successive crops. Until the recent introduction of seed-based fungicides, the inclusion of a break crop of a different species was the only practical way of keeping this disease under control.

Considerable genetic resistance to the major cereal diseases is to be found in wheat and barley, with the exception of take-all. Selection of resistant varieties is often the simplest and cheapest way of controlling diseases. Unfortunately other criteria may have an overriding influence on the selection process, for example, quality requirements or yield of grain. The emergence of new races of pathogens may also result in an established variety losing its ability to provide an accepted level of resistance against specific pathogens.

Despite the advances made in plant breeding and the adoption of appropriate husbandry strategies, control of the main cereal pathogens by non-chemical means are unlikely to provide total disease control. However, over reliance on chemical control measures is equally unacceptable as it can lead to the overuse of fungicides to the detriment of the environment and over time reduce the fungicidal properties of the chemical itself. For example, populations of *Erysiphe graminis* have developed resistance to some of the older mildewicides which has resulted in these products losing their effectiveness in controlling barley powdery mildew. The development of this kind of resistance has promoted the development of more integrated disease control strategies, where chemical control measures are combined with other measures to combat the pathogen.

10.7   Pest control

Cereals are susceptible to a wide range of pests. Some such as rabbits, birds and slugs attack a wide range of crop plants, whilst others such as wheat bulb fly, frit fly and cereal cyst nematode are specific to cereals. Plants at the seedling stage of growth are particularly vulnerable to pest damage when the green area of the crop is small. The established plant during the phase of active tillering is better able to compensate for moderate pest attacks, but crop damage can again be very significant after ear emergence. For the majority of crops, pest attacks are less threatening than disease outbreaks; therefore effective control strategies are based on early identification of the pest, an assessment of the population relative to the growth stage of the crop and the application of an appropriate pesticide. The aim must be to forestall epidemics at an early stage when a worthwhile cost benefit can be obtained from the control measure adopted.

Pest incidence in cereals varies according to crop species, soil type, farming practice and the weather pattern. Unlike diseases, variety resistance to pests is unlikely to be an important factor in the control strategy. An exception can be found in winter oats where certain varieties show resistance to stem eelworm (*Dtylenchus dipsaci*), whilst others are susceptible. Cereal cyst nematodes (*Heterodera avenae*) are frequently found to attack spring sown cereals sown on light, well drained soils often in areas where high quality malting barley is grown.

Intensive cereal production systems have often resulted in an increase in the incidence of pests, whilst the introduction of crops such as oilseed rape into the rotation may increase the incidence of slugs in the following cereal crops. There is some evidence that set-aside can provide a convenient green bridge which can aid the survival of some pest species between non-consecutive crops. Wheat bulb fly (*Delia coarctata*) may be more prevalent after set-aside if adults are able to lay their eggs on bare soil in the summer.

Weather patterns are known to influence the scale of future pest problems, for example high autumn rainfall preceding a mild winter is known to favour high populations of leatherjackets (*Tipula paludosa*) in spring. The incidence of barley yellow dwarf virus increases, especially in early drilled crops, if the autumn is warm, due to the increased activity of aphids.

Increasing emphasis is currently being placed on the production of high quality grain whilst minimising pesticide use. To achieve this objective more information is needed to identify situations where the use of pesticides cannot be avoided without serious financial loss. Forecasting pest activity by modelling their behaviour is one way in which more effective control strategies can be developed in the future. Threshold levels for individual pests, below which it is cost effective to apply a pesticide, are already helping to reduce pesticide usage. There is further scope to encourage beneficial organisms to counteract cereal pests in the field under the common aim of developing 'integrated pest management systems'. This approach places greater reliance on promoting vigorous crop growth through appropriate cultivations, rotations, and other agronomic inputs and the encouragement of beneficial organisms. In such a system selective pesticides could be used precisely timed to have the greatest impact on the pest, but to leave the 'beneficials' to prosper.

10.8   Harvesting and grain storage

Grain quality is largely determined during the growing season. Once the grain has been harvested it is difficult to improve its quality, although quality can easily be destroyed by conditions during harvest and subsequent drying and storage. Indeed grain quality can start to deteriorate in the field prior to harvest. Rainfall prior and during the harvest period can encourage ear diseases and premature sprouting, whilst high grain moisture levels will necessitate increased drying costs. For example, delaying the wheat harvest can result in grain with high α-amylase content, lower specific weight and protein contents, factors that will seriously reduce the quality of the grain for breadmaking.

Almost all UK grain will be harvested by large self propelled combine harvesters which are highly efficient, causing minimal physical damage to the grain when they are properly set and operated. The condition of the crop at harvest has a significant influence on combine performance; severely lodged crops not only produce inferior quality grain, but also reduce combine speed and efficiency. The presence of weeds also interferes with the harvesting process and often leads to higher grain moisture levels and contamination with weed seeds which increase the costs of cleaning. Unless the straw is required for animal feeding or bedding the combine harvester is extremely effective in chopping and spreading straw evenly over the soil. This is of considerable help in preparing the seedbed for the subsequent crop.

In Britain grain is often harvested at moisture contents of around 16 to 20%, whilst in exceptionally late seasons in northern regions grain may be harvested at around 25%. At these moisture levels stored grain is extremely susceptible to fungal contamination and deterioration. For safe storage the grain has to be dried to within the range 13 to 15% moisture, although malting barley may be dried to around 12%. Storage temperatures and grain moisture content have a strong influence on grain dormancy in barley, a condition which is highly important to the maltster.

Drying systems either rely on near ambient air temperatures, or high temperature driers. The former is a relatively slow process whereby the grain is stored in bins or on the floor and is dried by forcing ambient, or slightly warmer air through the grain. Batch and continuous flow high temperature driers rely on air temperatures of between 40 and 120°C with a necessity to cool the grain before storage. Drying temperatures of grain destined for milling, malting or for seed are more critical than for other uses.

10.9   References and further reading

AUSTIN R B (1978) 'Actual and potential yields of wheat and barley in the United Kingdom'. *ADAS Quarterly Review*, **29**, 76–87.

AUSTIN R B, BINGHAM J, BLACKWELL R D, EVANS L T, FORD M A, MORGAN C L and TAYLOR M (1980) 'Genetic improvement in winter wheat yields since 1900 and associated physiological changes'. *J. Agric Sci. Camb.*, **94**, 675–89.

GAIR R, JENKINS J E E and LESTER E (1987) *Cereal Pests and Diseases*. Farming Press, Ipswich.

GOODING M J and DAVIES W P (1997) *Wheat Production and Utilization, Systems, Quality and the Environment*. CAB International, Oxford.

HOME GROWN CEREAL AUTHORITY. *Weekly Market Information Bulletins*.

KENT N L (1982) *Technology of Cereals*. Pergamon Press, Oxford.

LUPTON F G H (ed.) (1987) *Wheat Breeding: its Scientific Basis*. Chapman and Hall, London.

NIAB (1999) *Cereal Variety Handbook*. NIAB, Cambridge.

NIX J (1999) *Farm Management Pocketbook*. Wye College, University of London, London.

PARRY D (1990) *Plant Pathology in Agriculture*. Cambridge University Press, Cambridge.

SILVEY V (1981) 'The contribution of new wheat, barley and oat varieties to increasing yield in England and Wales 1947–78'. *Journal of the NIAB*, **15**, 399–412.

SYLVESTER-BRADLEY R, SPINK J, FOULKES M J, BRYSON R J, SCOTT R K, STOKES D T, KING J A, PARISH D, PAVELEY N D and CLARE R W (2000) 'Sector challenge project – canopy management in practice' in *Proceedings of the HGCA Crop Management into the Millennium Conference*, Homerton College, Cambridge, January 2000.

WIBBERLEY E J (1989) *Cereal Husbandry*. Farming Press, Ipswich.

11

Summary and conclusions

P. C. Morris, G. H. Palmer and J. H. Bryce, Heriot-Watt University, Edinburgh

11.1 Improving cereal production and quality: a global challenge

The world faces the challenge of feeding, housing and clothing an ever-increasing population. On 12 October 1999, the United Nations marked the birth of the six billionth human currently living on the planet. This increase in population has been largely fuelled by medical advances which have allowed more children to survive infancy and reach child-bearing age themselves. The majority of these extra people live in the so-called developing world, where resources and cash are limited, and many live in poverty and hunger. Throughout our history there has been an overall increase in food production through agricultural innovations, the efforts of plant breeders and latterly great inputs of fertilisers and pesticides. But a continued, sustainable increase will be hard to realise. The task of maintaining the health and wellbeing of so many people without overburdening the already stretched natural resources of the planet are immense. Since, as we have seen in Chapter 1, cereals play such an integral part in global agriculture and diet (more than 50% of our food comes from three cereals: wheat, maize and rice), improved understanding of fundamental plant science and its application in plant breeding and technology will surely play an important role in the improvement of food production.

An important step forward in the feeding of the world was the green revolution of the 1960s and 1970s, during which, using conventional plant-breeding techniques, higher-yielding cereal varieties were developed and adopted by farmers in the developing world. International plant-breeding research institutes, sponsored and co-ordinated by the Consultative Group on International Agricultural Research, aided the development and dissemination of these improvements. However, there are limits to the improvements possible through conventional breeding methods, even with the aid of molecular biological techniques to monitor breeding programmes, since only those genes available in the breeding gene pool can be introduced. Additionally, the improved varieties developed during the green revolution have not always realised their full potential in the field because farmers have not had access to the appropriate fertilisers or pesticides.

11.2   Potential of cereal biotechnology

We have seen in Chapters 2 and 3 how it is possible to transform genetically the major cereals, such as wheat, barley, rice and maize, using techniques such as biolistics or direct DNA uptake into protoplasts. Since the information-encoding potential of DNA is a universally common factor between all living things, the theoretical and practical possibility exists to bring into crop plants genes from other plants or species that do not normally interbreed. DNA of choice can be manipulated by standard molecular biological techniques and introduced into plant tissue by a variety of methods such as protoplast-based technology, biolistics, or the use of the natural genetic engineering bacterium, *Agrobacterium tumefaciens*. This means that it should become possible to engineer plants that possess traits unimaginable through conventional breeding techniques, for example, plants with drought resistance, salinity resistance, virus resistance, improved yield and nutritional quality. In the highly industrialised economies, cereal plants adapted to specific processes and industries can be developed, for example, those with improved malting or baking qualities.

Even though the principles of plant genetic manipulation are relatively straightforward, this is still a fledgeling technology; it is not currently possible to manipulate genetically any established cereal crop plant at will. Many protocols, especially plant regeneration protocols, are very genotype dependent, that is, they will function only with certain varieties of a given crop. Once we understand the biology behind this selective regenerability, we may be able to extend the repertoire of transformable varieties. Very little work has been carried out on the genetic manipulation of crops such as millet and sorghum, of marginal importance to the West, but staples of Africa and India. Other technical bottlenecks to overcome include somaclonal variation, in which the regeneration process appears to be mutagenic and can lead to unwanted changes in the target plant, and transgene instability, where the introduced gene may no longer be expressed in later generations of the crop. A wider range of promoters to allow a greater choice of tissue-specific gene expression would also be a great boon to plant genetic engineers.

One very important application of biotechnology is in the use of molecular biological techniques in conventional breeding programmes, as detailed in Chapter 6. Established methodologies for cereal breeding have been, and remain, very effective in producing improved cereal varieties, but these methods are time consuming and therefore expensive. Increased efficiency in the identification of superior lines during a breeding programme by the use of DNA markers is a rapidly evolving area which promises to make an increasing contribution to the productivity of plant breeding.

11.3   Biotechnology in commercial practice

Chapter 3 illustrates how, from a commercial point of view, this young technology is being managed. For a plant biotechnology product to be commercially successful, many different objectives must be met. The market for the product must exist or be generated; this is a juxtaposition between 'technology push' and 'market pull'. The basic technology for the product must exist, and be available or be developed, and commercial lines must be developed, tested and marketed. These are lengthy processes and the costs are correspondingly high; lead-up costs must eventually be recouped in profits from the marketed product. Breeding and testing cycles can be hastened by transformation of elite commercial varieties (although as discussed above, this is not always possible), and by growing crops in different geographical locations that allow more than one generation per year. Importantly, risk assessment and regulatory issues must also be addressed (see also Chapter 7). And all of this must occur in a commercial environment with market-place competition and patented technology. Clearly, this is a capital intensive business and a great deal of planning and foresight must be employed.

The current and future ways in which transgenic technologies can be put to use to improve quantity and quality of cereals are outlined in Chapter 5. Current technologies in or near the market are relatively straightforward. The first generation of transgenic crops includes those with various forms of herbicide tolerance, which results in better weed control and thus higher yields of crops. Insect tolerance, for example by means of the insecticidal B.t. toxin, can also be built into crops, with the result that there is less insect damage through feeding and also indirectly since insects are important disease vectors. There is in addition the potential for less environmental damage since fewer sprays are needed for insect control. More sophisticated ways of managing plant disease based on endogenous or foreign genes are to be anticipated. We can also envisage improvements to the nutritional quality of cereals, for example an improved overall amino acid balance, which in maize, a staple component of the diet for many people, is relatively low in the essential amino acid lysine. Improvement geared to specific industrial processes such as brewing and baking can also be foreseen, such as alterations to the specific protein content of wheat to influence dough stickiness, or changes to starch-degrading enzymes so as to improve or alter the breakdown of reserve carbohydrates during malting and mashing.

11.4   Problems facing the cereal biotechnology industry

Although the potentials for cereal crop improvement through these technologies are enormous, there are also very real problems that must be overcome. A major problem that must be addressed is that of consumer acceptability, as discussed in Chapters 4 and 7. For a variety of reasons, the general public is somewhat sceptical about the adoption of transgenic plants, especially as foodstuffs. Even if many of these fears turn out to be unjustified, they are material and have already made a big difference to the marketing of these technologies.

Transformation technology currently relies on the use of selectable marker genes such as those encoding antibiotic resistances in order to select for transformed tissue during regeneration. The presence of these genes is undesirable. There is a very small possibility that ingested plant DNA can survive digestion, be taken up by gut bacteria and integrate into the bacterial genome in such a way that the bacteria become antibiotic resistant. From a scientific standpoint, this possibility is not thought to present a major hazard since antibiotic-resistance genes most often used for plant transformation, for example neomycin phosphotransferase, themselves come from bacteria; they encode resistance against antibiotics rarely used in human medicine (for example kanamycin) and a large percentage of gut microflora is already naturally resistant against kanamycin.

Antibiotic-resistant bacteria in a medical context are much in the news and any new developments which might produce more antibiotic-resistant bacteria, however small the risk or inconsequential the result, will not be acceptable in the market-place. Techniques are being developed that will enable selectable markers to be removed from crop plants after the transformation process. Alternative selectable markers not based on antibiotic selection are also being tested, for example a mannose permease that allows the use of mannose, a sugar not normally available to plant metabolism, as a carbon source during plant regeneration.

The potential presence of inadvertently introduced or produced toxins in transgenic plants must also be addressed, especially if the introduced genes code for proteins not normally found in the human food chain. In fact this approach has led to the discovery of the identity of an allergenic protein from brazil nuts. The gene encoding this protein was introduced into oilseed rape and bean plants in order to boost methionine content; routine allergy tests on the transgenic plants revealed the nature of the introduced protein.

Another fear is that of the spread of transgenes through pollen to related wild plants, which might cause, in the case of herbicide resistance, 'superweeds'. Most crop plants grown in Europe have few if any wild relatives that are sexually compatible. However, there have been reports of pollen, from oilseed rape for example, travelling many kilometres and 'weedy' oilseed rape can be an agronomic problem. A possible misconception in the mind of the public is that transgenic herbicide resistance confers resistance to all herbicides rather than one specific herbicide; nevertheless transfer of even one herbicide resistance gene would be undesirable. One potential method for avoiding spread of transgenes is to use the plant chloroplast as a target for transformation; pollen does not normally contain chloroplasts, and this would effectively limit the spread of a transgene outside of a breeding population. Chloroplast transformation technology is, however, still a developing science.

The potential negative effects of GM crops on biodiversity have also been raised. If crops have built-in pesticides, there may be a decrease in invertebrate fauna, and as a knock-on effect, less food for other animals. Similarly, herbicide-tolerant plants will result in more herbicide sprays and less plant biodiversity in cultivated fields, leading to a loss of animal habitat. Although it is difficult to envisage that this situation will be radically different from current industrial farming practice, the possibility of negative effects should be investigated and possible changes to alleviate the situation (also as a result of conventional agriculture) should be developed.

11.5   The future

Technology is science that works and the technology of milling, baking, malting, brewing and distilling do work, even if sometimes imperfectly. Biotechnology seeks to provide these industries with a raw material that will be closer to perfection. This will have the potential to increase the efficiency with which the same yield and quality of product can be achieved, or to improve the yield and/or quality of products. Thus both the producer and consumer should benefit from the application of biotechnology.

Current practices in milling and baking are described in Chapter 8 and in malting, brewing and distilling in Chapter 9. The following examples illustrate how every aspect of cereal structure, physiology, and biochemistry affect cereal processing. For example, the physical characteristics of grains affect their properties for milling (separation into germ, bran and endosperm), and their hydration during steeping at the start of malting. At a cellular/molecular level, the properties of proteins (gluten in wheat) and starch granules affect the suitability of the cereal for bread or biscuit production.

While structure is important, the physiology of the cereal will determine the environment in which it can grow successfully (Chapter 10) and whether it will malt successfully (lack of dormancy and an aleurone that is stimulated for the necessary time to produce appropriate levels of enzymes that can gain access throughout the endosperm). At the biochemical level, grains may need to produce a spectrum of enzymes if appropriate processing is to be possible. For example, the partial degradation of cell walls, protein, and starch in barley endosperm is required during malting and an appropriate enzyme profile is necessary later on in mashing to produce a fermentable wort. The ideal enzyme profile differs depending on whether beer or spirit is the final product.

The last fifty years have seen a massive increase in knowledge of the properties of the raw material in the food and drinks industries. Varieties of cereals have been developed to meet the needs of processors and innovative improvements in technology have assisted these processors in meeting the requirements of their customers. Over the years, cereal processors have developed a rapport with their suppliers of industrial equipment and this has enabled them to implement the most appropriate production technology. However, there has generally been no such rapport with cereal breeders.

The focus of breeding over the last 50 years has been to improve the arable yield. Using the natural genetic variation available, breeders have selected for those plants with the best agronomic traits including parameters such as disease resistance. From new varieties bred for yield, bodies representing cereal processors have then identified those varieties able to meet the specifications required by their industries. For example, trials are established to discover the extent to which new varieties are suitable for malting and the locations in which these varieties with malting potential can be grown. Lists of varieties deemed suitable for malting (approved varieties) are then provided to farmers.

At present, processors select the most appropriate varieties for their use. Over the next century, biotechnology will change this and varieties will be produced that more closely meet the requirements of the agronomist and processor. However, for this to occur, biotechnologists will have to understand the needs of the processor and the processor will have to grasp what can be achieved by the biotechnologist. This book has been written to initiate and enhance this dialogue.

Technology must now move from being the prerogative of the 'industrial plant' to include the 'biological plant'. One aspect of this change is the potential for direct genetic manipulation but this is only one aspect of how biotechnology will be used in the future. Genetic variation in plants provides an enormous untapped natural resource. Biotechnology will be used to bring this variation together in unique and useful ways and to identify characteristics of plants that can be utilised to enhance or extend their industrial performance.

Technology works in the specific circumstances in which it is applied. Biotechnology has to work at the appropriate time in a living organism that is developing at the molecular, biochemical, physiological and structural level while interacting with the environment (which may be far from a steady state). If biotechnology is to produce improved and new products, then wisdom, the possession of experience and knowledge together with the power to apply them critically or practically, must be pre-eminent.

Index